AuthorHouse™
1663 Liberty Drive
Bloomington, IN 47403
www.authorhouse.com
Phone: 1-800-839-8640

Published by AuthorHouse   6/12/2012

ISBN: 978-1-4772-1621-7 (sc)
ISBN: 978-1-4772-1620-0 (e)

INDEX

INTRODUCTION

In this authoritative text, the author delves into the nature of electromagnetic energy that attributes to the life force of the human being. Electromagnetic fields can be found in every animate and inanimate thing on earth, and can affect our mental state and general wellbeing.
This reference explores the sources, characteristics, health hazards and benefits of electromagnetic fields (EM), of low-frequency fields and of radio frequency radiations.

It focuses on bioengineering and biophysical aspects, including physiological and molecular genetic effects of time-varying electromagnetic fields on human neuronal cells. Some of the latest applications of EM fields, particularly for medical treatment and diagnostics are examined in this book. It contains materials describing different waves of EM of alpha, beta and Gamma to explain their interaction with biological systems.

The strength and location of natural magnetic fields on the earth can influence our minds, alter our character, and noticeably affect our health. Physicians have recognized the effect of magnetic fields of different altitudes on the earth. For example, the intensity of magnetic fields at the two poles is greater than at the equator.

This timely text discusses definitive reference for understanding the bioeffects of magnetism produced by DC and AC currents with static and with low and high frequencies. High voltage pulses and patients with bacterial and viral diseases are also discussed. The author of this text is a specialist in electrical engineering, who contributed to the field with a wide variety of interests and backgrounds.

Based on 2500 Hz being the best frequency to stimulate the muscle, the Neuromuscular Electrical Stimulation (NMES) causes the adaptation, i.e. training, of muscle fibers. Different types of pulses can activate different types of skeletal muscle fibers. The activation depends on the patterns of the NMES shape and duration. These patterns cause different response to different fiber types. Different patterns can be used for training, therapeutic, or cosmetic repairs such as muscle toning in the body, and micro-lifting of the face.

Electrical wave shapes passing through a nerve have also been demonstrated. The electrical wave shapes are caused by sodium and potassium, including the stimulation of slow and fast twitch muscle fibers.

The book covers defibrillation, which is a common treatment for life-threatening cardiac arrhythmias, ventricular fibrillation, and pulseless ventricular tachycardia. Cardiac arrhythmias happen when the heart beats too quickly, too slowly or with an irregular pattern.

General principles of somatosensory (auditory, visual, etc) evoked potentials, which are the electrical signals generated by the nervous system in response to sensory stimuli are also discussed. For evoked potentials, a pain stimulation using a helium laser stimulator with a laser-beam wavelength of 1.8 mm and a beam diameter of 30 mm^2 is used. Sometimes and other wavelengths and beam diameters are also exhibited.

The author handles the human disturbance of electrical interference, and their effect on DNA fragmentation. DNA fragmentation could lead to cancer. Cancer cells divide rapidly. The daughter cells divide before they have even reached maturity. Electromagnetic waves, pH, temperature, and some drugs (enzymes) may affect the rate of division.

This manuscript also deals with metabolism and ions emitted by electromagnetism. For example, in metabolism, the compound accepts or donates electrons in redox (reduction and oxidation) reactions. In metabolism, since NAD$^+$ is a cation, it is involved in redox reactions, taking electrons from one reaction to another. The coenzyme is therefore found in two forms in cells. The two forms are: NAD$^+$, which is an oxidizing agent and NADH, which is a reducing agent.

Electromagnetic fields can affect the aging factor. Oxidative stress due to radical components can damage mitochondrial DNA (mtDNA) and cell DNA. DC pulsed electromagnetic fields at under 10 Hz could enhance cell respiration mechanisms and lower free radical waste levels to a sufficient extent for the organism to perform as well as it did at 1/2 its age. Only sleeping in a 10 Hz field, however, will provide genetic effects as described by NASA. Above 10 Hz could damage all types of DNA and cell mitochondria. Electromagnetic radiation can be classified into ionizing radiation and non-ionizing radiation, based on whether it is capable of ionizing atoms and breaking chemical bonds. Both electromagnetic types can be associated with three major potential hazards: electrical, chemical and biological.

Pulsed Electro Magnetic Field Therapy (PEMF) is also expressed in pain relief, osteoporosis treatment, fibromyalgia treatment, fracture healing, rheumatic pain, and stress reduction.

The electrical system of the heart, as well as blood pressure, blood volume, and ECG in the heart (Wiggers Diagram) is included in the related chapter. The rhythmic contractions and expansions of the heart, which pump the blood occur in response to periodic electrical control pulse sequences. These together with cardiac systole and diastole are indicated with diagrams and detailed illustrations.

The electrical system of the brain is addressed here. The brain is represented by a closed box; of which the input can be one electrical pulse, and the output is of multi-order feed back loops. For example, you can measure the voltage, current, and frequency of the electrical signals input to the brain, but you can not measure simultaneous individual signals distributed among the parts of the brain. Each individual signal passes though complicated pathways to execute certain functions such as, perception, analysis, and task management. The voltage is determined primarily by the potassium and sodium ionic concentrations internal and external to the neuron. They are about -70mv at rest. Current flow is ionic, not electronic, and is not measurable, (from The Book of Intelligence and Brain Disorder" by this author). The book shows how the nerves work, including how they sendand receive messages through neurons in the body.

It is concluded that exposure to an electric field may benefit or damage certain parts of the body such as the liver, kidney, heart, red blood cells, muscle tissue and serum protein.

In the awareness of government demands for the outstanding safety of patients in hospitals, along with the rapidly expanding use of Magnetic Resonance Imaging (MRI), Nuclear Magnetic Resonance Imaging (NMRI), Magnetic Resonance Tomography (MRT) and medical imaging techniques, this book is particularly important. It is the definitive resource for information on the safety aspects of electromagnetic procedures.

ELECTRICITY IN HUMANS

Electricity and electromagnetism are forces of nature that can affect the life of beings either beneficially or harmfully, depending on their application. In medicine, electrotherapy can assist the healing process if correctly applied. However, the effects of electrical and magnetic fields have not been sufficiently studied regarding their role in the emergence of various disorders of the immune system like, AIDS, allergies, asthma, cancer, chronic fatigue, diabetes, fibromyalgia, lupus and other disease.

Electric and magnetic fields oscillate together but they are perpendicular to each other. The electromagnetic wave moves in a direction perpendicular to both of the fields, Figure (1I).

Figure (1I): Electromagnetic wave

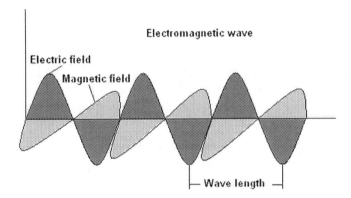

More over, electricity and electromagnetic fields do affect the spiritual quality of the human. Electricity can vibrate cells, nerves, bones, organs, and change the chemical reaction and metabolism. Disharmonious or unbalanced vibrations show themselves in many ways including discomfort, disease, illness, fatigue, and a variety of different symptoms. Higher vibrations mean healthy body, and low vibrations mean an ailing body.

Bioelectricity refers to the electrical, magnetic or electromagnetic fields produced by living cells, tissues or organisms. Bioelectricity includes the cell membrane potential and the electric currents that flow in nerves and muscles as a result of action potentials. An example of this the electric potential developed between sodium and potassium in the nerves.

For example, light is a magnetic wave and a part of quantum physics, and triggers melatonin hormone secretion. One of melatonin's physiological functions is to participate in the regulation of sleep onset and the quality of sleep. Melatonin is secreted in infinitesimal amounts by the pineal gland, which lies deep within the brain. At the end of the day, when our sunlight exposure decreases, melatonin begins to switch on, initiating relaxation and then sleep. People use melatonin supplementation to help them manage insomnia, regulate jet lag when travelling to different time zone, and to help them manage night shift work.

Electrical fields do not only affect the physical character of a body. They also influence the forces of soul and spirit-related power. In 1963 a pioneer in the field of electromedicine, the American physician Robert Becker, first formulated the theory that the natural, geographically determined magnetic environment presumably has an influence on human behaviour. It is known that the earth has a magnetic field concentrated at the two poles of the earth and is weakened in the centre (the equator). According to Rob Baker the human magnetic organ is situated in the bones of the apenoid and ethnoid which are directly located in front of the pituitary gland of the brain. The human magnetic organ feels the changes in the earth's magnetic field which is affected by environmental changes such as weather, earth quakes, tsunamis, etc. Any slight changes in the earth's magnetic field, even below 10 Hz, would disturb the apenoid/ethnoid and the pineal gland, which is a highly sensitive electromagnetic node in the brain. Such a disturbance would affect the consciousness, mode, emotion, Neurosemantic–Dexterity (awareness, prediction, calculation, etc) of a human being. The pineal gland reacts sensitively to very weak frequencies in the ELF (extremely low frequencies) which includes the natural micropulsation of a terrestrial magnetic field.

In 1960-1965, the US Embassy in Moscow was exposed to radiation from electromagnetic fields and microwaves created by the Soviets. As a result that various physical and mental disorders occurred among the American staff. The Soviets changed the atmospheric magnetic field to study the effect of terrestrial magnetic fields on humans. They revealed that "movement, emotion and behaviour can be controlled by electrical forces, and human beings can be controlled (José Delgado, Spanish neurologist). He summarized his findings by saying that "human beings can be controlled like robots with the touch of a button.

Two glands in the brain are exceptionally sensitive to electricity and light: The pineal gland and pituitary gland (shown in Figure (2)).

Figure (2): The location of the pineal gland and the pituitary gland in the brain

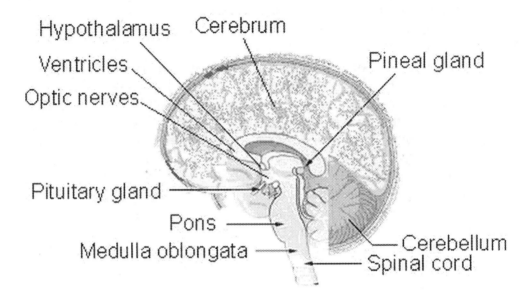

The two glands control thinking, sensory perception and procreation.

According to Rudolf Steiner's research in spiritual science, in the human being, behind the conscious sphere where memories are reflected, there is a subconscious region into which a substantial destructive focus has been introduced. In relation to the two cerebral glands it is of great importance that human recollective activity, that is the imprinting of the recollection, is connected to two etheric streams that are differentiated by the greatest possible tension. These streams have their physic-sensory counterpart in the heart region, the pineal gland and the pituitary gland. Steiner summarized his study in the following sentence: "the physical organ that wants to form the memory-image is the pineal gland, while the pituitary gland is the recording part".

So, the interaction between electricity and the human body is called "Bioelectricity" which is related to the electric phenomena in living organisms. Here are some examples:
- Bioelectricity enables migratory birds to travel thousands of miles at the same time each year with the accuracy that can only be detected by maps and GPS.
- Bioelectricity enables sharks to map out the ocean, and enables salmon to travel great distances and return back to their places of birth,
- Bioelectricity makes the electric eel generate large amplitude of electric voltage (up to 600 volts) and an electric field outside their bodies.

- Bioelectricity allows every cell in the body to pump ions through the change in concentration of Sodium-Potassium level (Sodium-Potassium pump). The ions are sent through nerves to all parts of the body.
- Bioelectricity charges cells with energy that produced by the metabolism of sugar, nitrogen and phosphor, is called "adenosine triphospate" (ATP). ATP can have neutral, negative or positive charges which interact with the magnetic fields of the earth, and further exemplify the human emotion and wellness.

In conclusion, the ancient Chinese described bioelectricity as "life-force". They call it "Qi" or "Chi" and they believed it permeated everything and linked their surroundings together. They represented it as the flow of energy around and in the body, forming a cohesive and functioning unit. Balancing the Qi in the human body can render wellbeing and happiness. By understanding its rhythm and flow they believed they could conduct exercises and treatments to provide stability, longevity and wellness.

Electroencephalography (EEG) is the recording of electrical waves along the scalp. EEG measures voltage fluctuations resulting from ionic current flows within the neurons of the brain. There are six wave patterns transmitted in the brain:

- The Delta pattern of 4 Hz with high amplitude, which is found in adults and babies.
- The Theta pattern with frequency ranges between 4 Hz and 7 Hz, found in children and in meditation.
- The Alpha pattern with frequency ranges between 8 Hz and 12 Hz, found in the posterior regions of the head on both sides. It can be seen by closing the eyes and with relaxation, and attenuates with eye opening or mental exertion.
- The Beta Pattern with frequency ranges between 12 Hz and 30 Hz. It is seen usually on both sides in symmetrical distribution and is most evident in the frontal lobe. Beta activity is closely linked to motor behavior and is generally attenuated during practicing movements.
- The Gamma pattern with frequency ranges between 30 Hz and 100 Hz, are thought to represent the binding of different populations of neurons together into a network for the purpose of carrying out a certain cognitive or motor function.
- The Mu pattern is similar to the Alpha pattern, with frequency ranges between 8 Hz and 13 Hz. It reflects the synchronous firing of motor neurons in the rest state. Figure (3) shows the six patterns.

Figure (3): Waves of Electroencephalography (EEG)

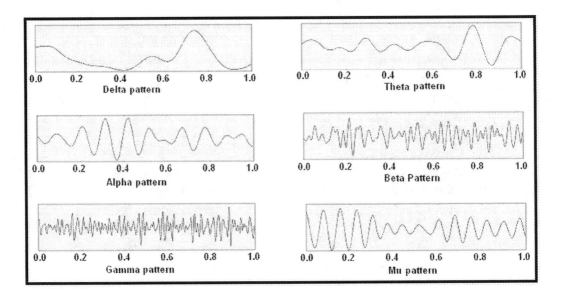

Chapter 1 - How Does The Body Make Electricity?

The human body generates electricity in two ways:

a) The Panasonic Company is looking into using human blood to generate electricity to power electrical devices. It's looking into how blood could break down sugars to generate electricity, through the Krebs (citric) cycle, like it generates energy for the human body. The produced electricity from the blood could be used to power nano robots and nano-devices implanted in the body. These devices include nano robots and nano devices with multiple degrees of freedom, which have the ability to apply biological functions, and navigate and move through our bodies. Taking inspiration from the biological motors of living cells, chemists are learning how to utilize protein dynamics to power microsize and nanosize machines with catalytic reactions.

b) In the natural resting state of your body, the neuron cells are electrically stable, although they are negatively charged. Negativity won't have any real intensity and it won't be capable of overwhelming, or holding, your being anymore. Negativity yields from the potassium inside the neuron cell which has sodium on the outside. Potassium is more negative that sodium (K has 2, 8, 8, 1 electrons, and Na has 2, 8, 1). Each cell has a gate which opens and closes when an imbalance between negativity (potassium inside the cell) and positivity (sodium outside the cell) is created. For example, anger increases the sodium positive ions to open the gate and enters the cell to mix with the potassium. These result in sending positive impulses through the nerves to the heart and brain (see the function of the heart in the coming chapter). If you are happy, a negative impulse will flow in the nerves. The protein of the neuron loves the potassium and hates the sodium, thus, the neuron pumps out the sodium and pumps in the potassium. Because the sodium atom ends with one electron and the chlorine in the sodium chloride (salt) ends with seven electrons, the sodium atom looses the electron to the chlorine atom. Thus the sodium atom becomes a positive ion (cation), and the chlorine atom becomes negative ion (anion). Balanced electrolytes: the lack of sodium and potassium electrolytes can reduce critical brain functions. The balance of electrolytes is also important with potassium being needed in larger quantities.

Figure (1.1) shows the mechanism of the sodium and potassium electrical pump.

Figure (1.1): The mechanism of the sodium and potassium electrical pump

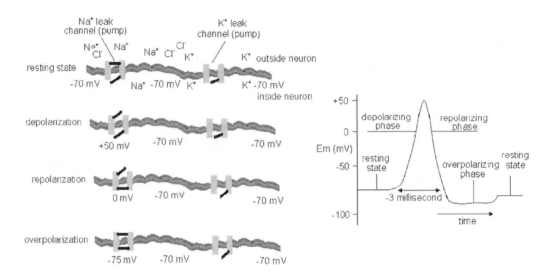

The impulses are sent to the brain, which in turn modulates substances that act as neurotransmitters. they can be broadly classified into three major groups:

1. Amino acids such as glumatic acid (glutamate), GABA, aspartic acid and glycine
2. Peptides such as vasopression, somatostatin, and neurotension
3. Monoamines such as neropinophrene, dopamine serotonin and acetylcholine

The central machine of the brain's neurotransmitters is made of glutamate and GABA. Some examples of the action of the impulses on the brain's neurotransmitter are:

- Dopamine – voluntary movement
- Serotonin – sleep and temperature regulation
- GABA (gamma aminobutryic acid) – motor behaviour
- Glycine – spinal reflexes and motor behavoiur
- Noradrenaline – wakening /alertness and arousal
- Acetycholine – voluntary movement of the muscles
- Neuromodulator – sensory transmission (pain)
- Enkephalin (opiate) – stress, pain killer, promote calm
- ATP – energy
- Insulin – sugar

Chapter 2 – How Humans Conduct Electricity

The human body is a good conductor of electricity as it is mostly water, plus salts, minerals and a whole bunch of other stuff that makes that water very conductive to electricity. There are millions of charged ions all over the body, and when an electric current or magnetic field is applied, they line up in the direction of the field. The salt water in the human body including the skin is about 70% of the weight. Salt water is a good conductor of electricity. The neurological system is also a good conductor.

There must be a resistance or impedance in order to have a voltage potential between two points. In the resistance, the voltage is in parallel with the current, whereas the current lags the voltage in the inductance, and leads in the capacitance, as seen in Figure (2.1).

Figure (2.1): Current and voltage in resistance, inductance and capacitance

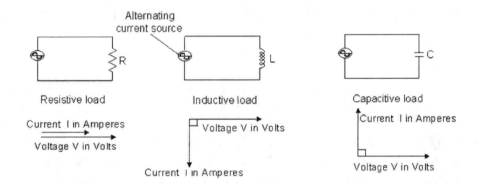

The human body contains many forms of capacitive loads due to the different elements in the body. Bones, teeth, hair, fat, cartilages, etc. contribute to capacitive loads. DNA is the main inductive element. Capacitive loads are used as storage to electrical fields, which are discharged into other loads. Thus, static electricity can be produced when feet are dragged on non metallic surfaces (a nonmetallic surface is considered to be a dielectric means of a capacitor) or ungrounded metallic surfaces.

The nervous system is composed of two parts: the central nervous system, which comprises the brain and the spinal cord, and the peripheral nervous system, which consists of nerves connecting other segments of the body to the brain through the spinal cord. The nervous system is very complicated, and the designer of such a system must have an intricate intelligence in harnessing electrical energy to flow through the nerve system and control all functions of the human body. It is not an undirected process like natural selection, which is alleged to be the key mechanism of evolution. When we ponder the complexity and the operation of the nerve system, we can only wonder at Gods overpowering creation and wisdom.

The nervous system is filled with references to electrical theories and electrical concepts that science uses today. Such references include technical words like circuit, network, current, charge, power, potential, battery, resistance, inductance, alternating, impedance, signal, noise, frequency, electromagnetic, motor, generator, transmission, microwave, etc. The difficulty of describing the nervous system without resorting to such language implies the God understanding prior to man's electrical discoveries.

2.1 The Nervous System

The human nervous system consists of two main systems: the central nervous system (CNS), and the peripheral nervous system (PNS). This includes the somatic motor nervous system, and the sensory nervous system. The central nervous system contains the majority of the nervous system and consists of the brain and the spinal cord. The main function of the PNS is to connect the CNS to the limbs and organs. The peripheral nervous system is divided into the somatic nervous system and the autonomic nervous system.

2.1.1 Peripheral Nervous System (PNS)

The neuron consists of two portions: the cell body and the axon. The cell body is like the other cells; it contains a nucleus and cytoplasm. It is also different from other cells because (long threadlike projections protrude) out of the cell body. These are called dendrites ("tree" in Greek). At one point of the cell, however, there is a particularly long extension that usually does not branch throughout most of its length. This is the axon (the axis), Figure (2.2).

Figure (2.2): Neuron

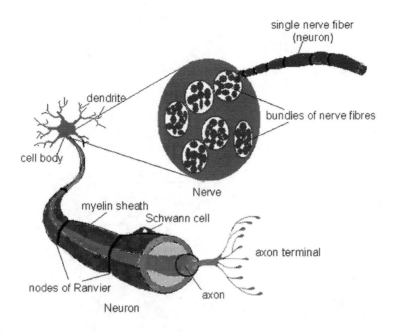

Structurally, a neuron consists of dendrites, an axon and an axon terminal. Dendrites conduct nerve impulses toward the cell body. The axon conducts nerve impulses away from the cell body through the axon terminal. To speed up the transmission and to keep the signal from scattering and propagation (like an electrical cable), the axon is sheathed with a myelin layer that is made up of Schwann cells. Messages from neurons move as fast as 400 km per hour.

The axon terminal receives messages from the cell body, and then transmits the messages to neighboring neurons via the release of neurotransmitters, Neurotransmitters are endogenous chemicals that relay, amplify, and modulate signals between neurons and other cells, Figure (2.3).

Figure (2.3): Sending and receiving messages through neurons

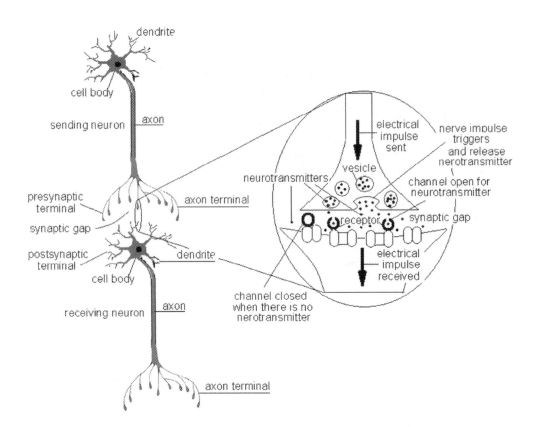

2.1.2 Types of Neurons

The three basic types of neurons are the motor neuron (efferent), the sensory neuron (afferent), and the interneuron. The motor neurons are specified to send messages away from the Central Nervous System. The sensory neurons specify in the senses of taste, touch, hearing, smell, and sight. They send messages from the sensory receptors to the Central Nervous System. The interneurons are basically a mix of both a sensory neuron and a motor neuron, Figure (2.4).

Figure (2.4): Sending messages to and from the Central Nervous System

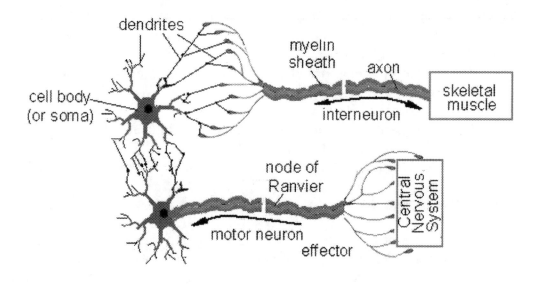

2.1.3 Transmitting Messages into the Neuron

At rest, there is an electrical charge difference between the inside and outside of the neuron because of either positively or negatively charged ions that are caused by sodium (Na^+), potassium (K^+) and chloride (Cl^-). The inside of the neuron is more negatively charged than the outside of the neuron (because sodium is more than ten times more concentrated outside the neuron's membrane than inside of the neuron), and the neuron is said to be polarized, i.e., there is a difference in electrical charge between the inside and outside of the neuron.

The neuron has channels that can permit chemicals to pass in and out of it. The sodium channels are completely closed during the resting potential, but the potassium channels are partly open, so potassium can flow slowly out of the neuron.

The protein of the neuron loves the potassium and hates the sodium. Thus, the neuron pumps out the sodium and pumps in the potassium. Because the sodium atom ends with one electron and the chlorine in the sodium chloride (salt) ends with seven electrons, the sodium atom loses the electron to the chlorine atom. The sodium atom becomes a positive ion (cation), and the chlorine atom becomes negative ion (anion).

2.1.4 The Central Nervous System (CNS)

The Central Nervous System (CNS) consists of:

- The brain which contains:
  1. The Rhombencephalon or hindbrain which consists of medulla, pons and cerebellum
  2. The Mesencephalon or midbrain
  3. The Prosencephalon or forebrain which consists of the diencephalon and the telencephalon
  4. The Cerebral hemisphere which consists of many smaller parts

- The spinal cord

  The peripheral nervous system (PNS) is composed of nerves and ganglia. A ganglion is a collection of neuronal cell bodies outside the CNS. Typically a ganglion is a lump on a nerve, but many of the ganglia associated with internal organs are of microscopic size. The peripheral nervous system is made up of all neurons in the body outside of the central nervous system, and includes:

  1. The sensory somatic nervous system is made up of afferent neurons that convey sensory information from the sense organs to the brain and spinal cord, and efferent neurons that carry motor instructions to the muscles. The sensory-somatic system consists of 12 pairs of cranial nerves and 31 pairs of spinal nerves.

  2. The autonomic nervous system consists of sensory neurons and motor neurons that run between the central nervous system (especially the hypothalamus and medulla oblongata) and various internal organs such as: the heart, lungs, viscera and both exocrine and endocrine glands.

  The autonomic system has two groups:

  a- The sympathetic nervous system activates what has been called the "fight-or-flight" response that prepares the body for action
  b- The parasympathetic nervous system conserves energy and prepares the body to rest.
  The diagram of the human nervous system is shown in Figure (2.5).

Figure (2.5): Human nervous system

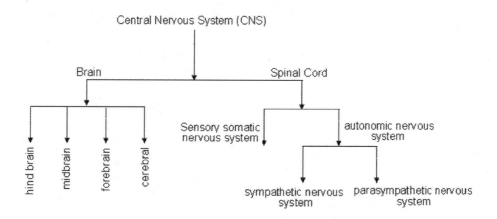

All of the pieces of information decoded in the various sensory areas of the cortex converge in the hippocampus, which then sends them back where they came from. Figure (2.6) shows the connection between the nerve system and body's parts.

Sensory information from the physical world to our brain is relayed via our five senses, using electrical switches that change one form of energy into electrical energy. Our bodies have sensory receptor cells because there are different types of physical stimuli to be changed into electrical signals. For example, a different type of receptor cell is required for hearing stimuli than for smelling stimuli.

Figure (2.6): The nervous system

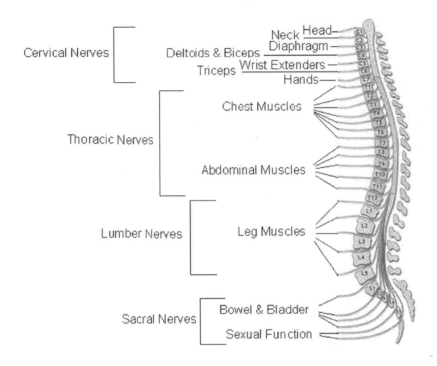

2.2 The Electrical Pathway for Fear and Pain

2.2.1 Electrical Pathway of Hormones associated with Fear

The brain is constantly transferring information and triggering responses through electrical pulses and hormones. There are many parts in the brain at least peripherally involved in fear. Research has discovered that certain parts of the brain play central roles in the process. Whenever we perceive a risk that we are physically or psychologically threatened, an inherent, reflex, alarm-system in our brain triggers the release of electrical impulses and a variety of hormones. There is an intricate group of over 30 stress hormones, such as adrenaline, noradrenaline epinephrine, norepinephrine and cortisol, that have a powerful and widespread effect on our body's biochemistry, physiology and psychology. Hormones and electrical pulses cause changes in the body that include:

- Heart rate and blood pressure increase
- Pupils in the eyes dilate to take in as much light as possible
- Veins in the skin constrict to send more blood to major muscle groups (responsible for the "chill" sometimes associated with fear -- less blood in the skin to keep it warm)
- Blood-glucose level increases

- Muscles tense up, energized by adrenaline and glucose (responsible for goose bumps -- when tiny muscles attached to each hair on surface of skin tense up, the hairs are forced upright, pulling skin with them)
- Smooth muscle relaxes in order to allow more oxygen into the lungs
- Nonessential systems (like digestion and the immune system) shut down to allow more energy for emergency functions

The brain and the body will be in a state of fight and flight. If a wild animal attack us, we can either escape (flight) or fight.

The process of transforming the data fromfear is shown in Figure (2.7).

Figure (2.7): The activation of the "fight or flight" response

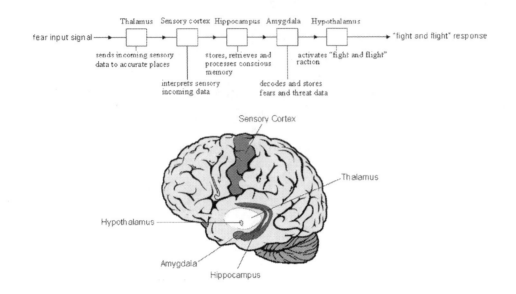

2.2.2 The Electrical Pathway and Hormones with Pain

The International Association for the study of pain classified pain as an "unpleasant sensory and emotional experience associated with actual or potential tissue damage." This meaning is pretty ambiguous and indefinite. Pain is the process by which a painful stimulus is transmitted from the site of pain to the central nervous system. Scientists define some ways inwhich the pain perception is transmitted to the brain. The stimuli can be mechanical (cuts, punctures, pressure, stubbing a toe, burning a finger) or chemical (putting alcohol on a cut, eating poisonous food. The ending nerves (dendrites) send the signal to the central nervous system (the brain) for further processing and action.

Both inflammation and pain are associated with a number of diseases. It has been proposed that substance P (or calcitonin gene-related peptide [CGRP]), released from primary afferent nerve endings play a role in these conditions. Substance P (SP), also released in other nodes in the peripheral nerves, is a neuropeptide (protein) which functions as a neurotransmitter and as a neuromodulator. Recent developments show that scientists have demonstrated the importance of substance P in several models of disease. A role for SP is proposed in the regulation of pain, asthma, psoriasis, inflammatory bowel disease and, in the CNS, emesis, migraine, schizophrenia, depression and anxiety. Once the pain information is in the brain (we're not quite sure how it gets processed), some signals go to the motor cortex, then down through the spinal cord and to the motor nerves.

These impulses would cause muscle contractions to move your hand out of the way of whatever is causing the pain. Figure (2.8) shows the peripheral nerves and the substance P relayed to the brain.

Figure (2.8): Substance P is released and relayed to the brain though peripheral nerves

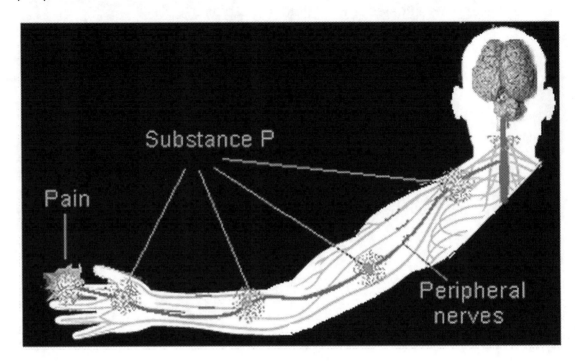

2.3 Human Body Organs that Can Produce an Electric Current

There are many organs that can produce electricity:

1- Nerves

The nerves can produce an electric current from pulses. An electric current or pulse can be transmitted to the brain. Neurons are not good conductors for electricity (like copper or aluminum), however, they can transmit spikes or pulses, which are usually characterized by high voltage and high frequency. The voltage in neurons at rest is negative and reaches about 50-70 millivolts. Spikes and pulses can reach 1-2 volts (this is a high voltage considering the voltage of human bodies). In industrial electrical transmission lines, high voltage can reach 600, 000 volts and more, where pulses and impulses can reach millions of volts. Electromyography (EMG) is a technique for evaluating and recording the electrical activity produced by nerves. EMG can be used to indicate if there is damage in the nerves or muscles.

Electrical currents in wires can be reduced by using resistances and impedances (combination of resistance, inductance and capacitance), or can be stopped from flowing if the electrical circuit is opened or cut. On the other hand, electrical current yields from pulses can also be inhibited by using anesthetics.

In real life, electrical pulses produce heat as they travel along wires, but "electrical pulses in nerves produce sound measured in few milli decibels (dBm)," says associate professor Thomas Heimburg from the Neils Bohr Institute at Copenhagen University. Dr. Heimburg explained that the sound is converted to a mechanical force or vibration in the body.

Recent research suggested that anesthetics like ether, laughing gas, chloroform, procaine and the noble gas xenon can freeze the fatty layer (myelin) which cover the nerves and block the sound. Thus the nerves would be "turned off". In spite of this, no one knows precisely how anesthetics work. How are the nerves "turned off"? Nerve stimulus triggers an electric discharge into the nerve which propagates down the fiber (axon) until it reaches the axon terminal bundle on the order of tens of meters per second. The discharge will continuously be transmitted to a neighboring nerve cell.

The nerve cell is similar to a receiver, transmitter and transmission line, passing the signal from the dendrites to the axon terminal. When the signal reaches the axon terminal, it proceeds down to the next nerve cell by successive excitation of the segments of the axon membrane. The signal is successively stimulated by the effect of sodium and potassium in the Nodes of Ranveir, Figure (2.9). The myelin sheath around the axon prevents the signal from escaping or becoming distracted. The myelin sheaths are about 1mm in length. The action potential travels along the axon at speeds from 1 to 100 m/s.

Figure (2.9): Successive stimulation of a pulse

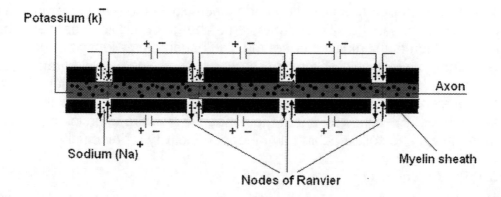

2- Muscles

Muscles receive the current produced by the nerves and transmit it to the brain. The voltage in the muscle ranges from 6-12 mV negative depending on the muscle type, and the frequency between 7-25 Hz. For example, the eye motor muscle has a voltage of about 6 mV, and the buttock (gluteal) muscle has one of 12 mV. When the muscle is damaged, the voltage can reach to 800 mV. Some animals have muscles that can produce 600 V; enough to cause stunning or killing of people and fish. The electric eel is one of the few animals on the planet that can make, store, and discharge electricity. Like a battery, the electric eel has two opposite poles (the head and the tail), and when they discharge, the voltage flows from either the head or the tail. The eel is vey aggressive when the head is positive. The shock from an electric eel disturbs the body by altering physiological functions causing involuntary muscle actions and respiration. Symptoms of being shocked by an electric eel can be respiratory paralysis and cardiac failure. These symptoms may result in death.

According to Steve Clinefelter, President of California Fitness, "one person has the ability of producing 50 watts of electricity per hour when exercising at a moderate pace. If a person spends one hour per day running on the machine, he/she could generate 18.2 kilowatts of electricity and prevent 4,380 liters of CO_2 released per year." The technique of the production of electricity from muscles is controversial and not yet known.

3- Skin

Skin can also produce an electric current, often referred to as the galvanic skin response. There are only a couple of places where it is widely recognized as easy and reliable to measure the electricity: the palms and the soles of the feet. In these places there is a high density of the sweat glands, which are known to be responsive to emotional and other psychological stimuli. In either of these locations, the electricity is measured by placing two microelectrodes in the skin and passing a charge of millivolts between the two points. When the stimuli increase, the skin becomes a slightly better conductor of electricity. This response can then be measured by an AMG or a voltmeter of high sensitivity. All skin glands have different conductivities to the flow of electrical currents. The armpits and genital glands have little conductivity. Exercise can affect the conductivity. More exercise produces more sweat and salt, reaching a higher conductivity for electricity.

3.1 The Galvactivator

The galvactivator was designed by Jocelyn Scheirer and Rosalind Picard of the MIT Media Laboratory. Elements of the circuitry were implemented by Dana Kirsch and Blake Brasher, also of the Media Lab. The glove itself was designed in collaboration with Philips Electronics. The galvactivator can sense the wearer's skin conductivity and then it maps its values to a bright LED display, making the skin conductivity level visible. Increases in skin conductivity tend to be good indicators of physiological arousal --- causing the galvactivator display to glow brightly, Figure (2.10).

Figure (2.10): The Galvactivator

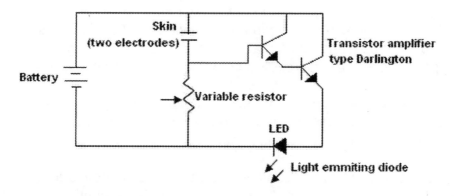

2.4 Brains-Signal Simulation

The human brain has a very complex way of sending and receiving signals of actions and movements. Each signal has a start and stop point. The researchers explored the role of brain circuits located in the basal ganglia, and found that dopamine-producing neurons project into the striatum and terminate at the substantia nigra. The substantia nigra is located in the basal ganglia, and both play an important role in the initiation and termination of newly learnt behavioural sequences. Substantia nigra means "black substance" in Latin. It appears darker than neighboring areas due to high levels of melanin in the neurotransmitters (dopamine). The basal ganglia contains the striatum, globus pallidus and subthalamus nucleus.

Rui Costa and Xin Jin (National Institutes of Health,USA) show that when mice are learning to perform a particular behavioural sequence there is a specific neural activity represented in electrical signals that emerges in their brain circuits and signals the initiation and termination steps. Interestingly these are the circuits that degenerate in patients suffering from Parkinson's and Huntington's diseases, who also display impairments both in sequence learning, and in the initiation and termination of voluntary movements.

Xin Jun adds: "This start/stop activity appears during learning and disrupting it genetically severely impairs the learning of new action sequences. These findings may provide a possible insight into the mechanism underlying the sequence learning and execution impairments observed in Parkinson's and

Huntington's patients who have lost basal ganglia neurons which may be important in generating initiation and termination activity in their brain".

The network of brain circuitries is too complex to describe, but molecular biology and computing methods have improved to the point that the National Institutes of Health have announced a $30 million plan to map the human "internal connection and signal pathways".

The University of Southern California neuroscientists Richard H. Thompson and Larry W. Swanson used the method to trace circuits running through a "hedonic hot spot" related to food enjoyment. The circuits showed up as patterns of circular loops, suggesting that at least in this part of the rat brain, the wiring diagram looks like a distributed network of the Internet.

"We started in one place and looked at the connections. It led into a very complicated series of loops and circuits. It's not an organizational chart. There's no top and bottom to it," Swanson said.

The author of this book (Dr. Amin Elsersawi) believes that incoming and outgoing signals from the basal ganglia to each part of the brain are similar to a spider web of multi-feed back loops. If you knock out any single "thread" of the web the rest of it works. He agreed with Swanson when she said, "There are usually alternate pathways through the nervous system. It's very hard to say that any one part is absolutely essential."

Dr. Elsersawi suggested an imaginative electrical diagram representing the brain network as depicted in Figure (2.11).

Figure (2.11): Signal pathways in the brain

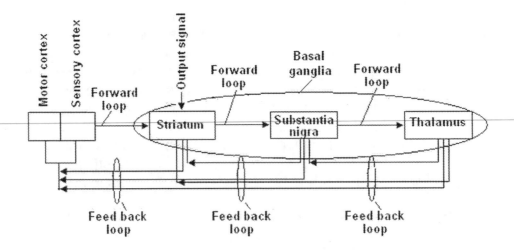

Chapter 3- Static Electricity

The term static refers to a situation where the fields do not change with time. Static electric and magnetic fields are characterized by constant amplitude and frequency of zero Hz. They are similar to the DC current because they have a steady flow in one direction and a steady rate, unless there are changes in the surrounding media.

3.1 Capacitance of the Human Body

Any object that would normally be considered an electrical insulator, i.e. resisting the flow of electricity, can have a buildup of natural electrical charges on its surface. These are called static charges, or static electricity, which remain until they are discharged by contacting a conductor or contacting the ground.

The human body can be considered as a capacitor, depending on many factors such as environmental media, the path of electricity through the body, the fat in the body, the sweat and salt accumulated on the skin, etc. There are a lot of factors involved and not every person has the same electrical capacitance and resistance. For instance, men tend to have lower resistance than women. The resistance of a person's arm depends on the arm's length and diameter. Resistance goes up with length and down with diameter (R ∞ L, ∞ 1/A, L = length, A = area). Since men tend to have thicker arms and legs (more muscle), they usually have lower resistance and capacitance. This means that lightning discharge in men is higher than in women. A rough value for the internal resistance of the human body is 300-1,000 Ohms, and the capacitance in normal surroundings is typically in the tens to low hundreds of picoFarads. The amount of stored energy is relatively low, and won't harm a healthy person. Naturally, the resistance also depends on the path that electricity takes through the body - if the electricity goes in the right hand and out the left foot, then the resistance will be much higher than if it goes in and out of adjacent fingers.

The energy stored as static electricity on a human body depends on its capacitance, the voltage to which is charged, and the dielectric constant of the surrounding medium. A capacitor consists of at least two electrical foils (conductors) separated by a dielectric insulator. Capacitance is measured in Farad.

Studies on cells or cellular components are useful for understanding interaction mechanisms of static magnet fields. They are not sufficient to identify health effects, but can give an indication of the sorts of effects that might be investigated in animals and humans. Static electric fields generate a surface electric charge and are not appropriately studied in vitro.

An intriguing claim that human brain cells possess crystals of a highly magnetic mineral known as magnetite is described by Dr. Joseph Kirschvink, a professor at the California Institute of Technology.

The 38-year-old geobiologist says that he believes that magnetite crystals enable animals from bees to whales to navigate by using the earth's magnetic field. He says he doubts that they support any sensory capability in humans, although he suspects that they might account for the possible influence of strong electromagnetic fields on human health.

For modeling the effect of static discharge on sensitive electronic devices, a human being is represented as a capacitor of 300 picoFarads, charged to a voltage of 5000 volts, Figure (3.1). When touching an object this energy is discharged in less than a microsecond. While the total energy is small, on the order of milliJoules, it can still damage sensitive electronic devices. Larger objects will store more energy, which may be directly hazardous to human contact or which may give a spark that can ignite flammable gas or dust. The discharged energy is calculated by using the formula $E = \frac{1}{2} CV^2$. IEC-2:1987 states that a discharge with energy greater than 5000 mJ is a direct serious risk to human health. IEC60065 states that consumer products cannot discharge more than 350 mJ into a person.

Figure (3.1): Model of a lightning strike to the human body

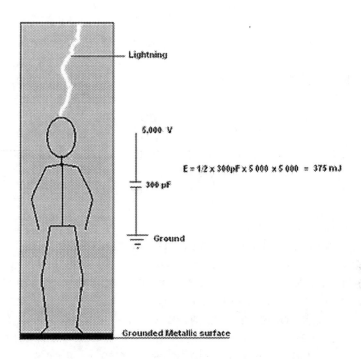

So, if the capacitance of the human body is as high as 300 PicoFarads, and has a charge of 5000 volts, discharged, e.g. during touching a charged car, will create a spark with the energy of 375 millijoules. The maximum potential commonly achieved on a human body ranges between 1 and 10 kV. The maximum potential is limited to about 35–40 kV, due to corona discharge dissipating the charge at higher potentials, http://en.wikipedia.org/wiki/Static_electricity#cite_note-21#cite_note-21.

As little as 0.2 millijoules may present an ignition hazard; such low spark energy is often below the threshold of human visual and auditory perception.

3.2 Static Electric Fields and Magnetic Fields

Electric and magnetic fields (EMFs) are invisible lines of force associated with the generation, transmission, and distribution of electric power such as those associated with transformers, high-voltage transmission lines, conversion from ac to dc and vise versa, secondary power lines, and home power and lighting. EMFs can also be generated by natural phenomena like the Earth's Magnetic field, lightning and a solar blaze (harsh geomagnetic storm released from the sun, which could disrupt power grids, radio infrastructure, and GPS as well as flash stunning auroras). Electric and magnetic fields also arise from the motors and heating coils found in electronic equipment and appliances.

An electric field is the force field created by the attraction and repulsion of electricity discharged and is measured in volts per meter (V/m). A magnetic field is created by a magnet or as a consequence of the movement of charges (flow of electricity). It is measured in Tesla.

Static electric fields occur naturally in the atmosphere due to lightning strikes, friction (for example walking on a carpet) and the rotation of metallic conductors in magnetic fields (motors and generators).

Static magnetic fields are produced by moving electric charges. The intrinsic magnetic moments of elementary particles generate electric fields and currents. They can be discharged to and around the heart, and slightly impede the flow of blood, affect metallic implants like pace makers, and interfere with some chemical reaction in the body.

A number of different biological effects of static magnetic fields have been explored, by studying cells or their components and processes. These include changes in cellular processes, gene expression, cell growth and genetic material. The findings have been contradictory. Several effects have been observed at field strengths lower than 1000 mT, but most results were not replicated by other researchers. Overall, the experiments on cells conducted so far do not present a clear picture of specific effects of static magnetic fields and do not indicate possible adverse health effects.

Studies focusing on the nervous system consistently indicate that laboratory rodents feel discomfort when moving in static magnetic fields of 4000 mT or more and try to avoid them. This is thought to be linked to effects of magnetic fields on the system in the inner ear which controls balance and body orientation.

3.3 The Environmental Effect on Static Electricity

Generating the spark depends on how much charge you have built up and the environment e.g. humidity levels. In a low humidity and clean room, static charge is built up more efficiently than in a dry room. Here are some factors that affect static electricity:

Humidity

In heated buildings, low humidity (less moisture in the air) is common. Static electrical discharge is more common in low-humidity environments and can damage electronic components. A person who touches the equipments is exposed to a dangerous electrical shock. That is why a ground fault circuit interrupter (GFCI), which is an electrical device that protects personnel by detecting potentially hazardous ground faults and quickly disconnecting power from the circuit, must be installed. It's not the voltage but the current that kills. People have been killed by even less than 100 volts AC in the home and with as little as 40 volts DC. The shock's intensity depends on the amount of current passed into the body. Any electrical device used on a house wiring circuit, if not properly grounded, can carry a fatal amount of current. Current capacities between 100 and 200 milliamperes (0.1 ampere and 0.2 ampere) are fatal. Anything in the neighborhood of 10 milliamperes (0.01) is capable of producing painful to severe shock, Table (3.1). The bathroom in a normal house is the most dangerous room. Nearly every year someone is electrocuted in a bathroom because of violations in the wiring. Generally, about 1000 people die in USA every year because of electric shock.

Table (3.1) shows the effect of different current capacities.

Safety	Current	Effect
Safe	1 mA or less	Causes no sensation
Safe	1 mA to 8 mA	Sensation of shock, not painful; Individual can let go at will since muscular control is not lost.
Unsafe	8 mA to 15 mA	Painful shock; individual can let go at will since muscular control is not lost.
Unsafe	15 mA to 20 mA	Painful shock; control of adjacent muscles lost; victim can not let go.
Dangerous Shock	50 mA to 100 mA	Ventricular fibrillation - a heart condition that can result in death - is possible.
Dangerous Shock	100 mA to 200 mA	Ventricular fibrillation occurs.
Dangerous Shock	200 mA and over	Severe burns, severe muscular contractions - so severe that chest muscles clamp the heart and stop it for the duration of the shock. (This prevents ventricular fibrillation). Chest muscles clamp the heart and stop it for the duration of the shock. (This prevents ventricular fibrillation).

The effect of the electrical shock depends on two factors: the amplitude of the current, and the duration of the shock. Figure (3.2) shows how the duration of the shock affects the hazard of ventricular fibrillation for 60-cycle alternating current.

Figure (3.2): Effect of amplitude of current and duration of the shock on health

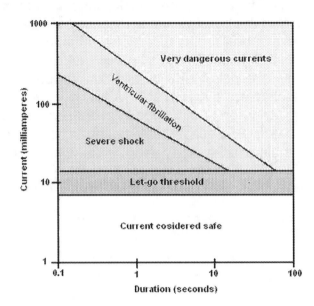

For safety, read or consult OSHA Standard 29CFRI910.147

Dust and Smoke:

Dust and smoke can absorb moisture, which may then result in a short circuit or excessive current leakage to the ground. If wires are exposed, they may cause a shock to a person who contacts them. Excessive dust on electronic equipment can cause electrical shorts and disastrous equipment failure.

Smoke causes a kind of contamination that is similar to the effects of dust. It coats the surfaces of electronic components, acting as both an insulator and a conductor, depending on the type of the smoke. In electronic devices (computers), dust buildup in their cooling passages can be a problem because the dust acts as a thermal insulator, which prevents proper cooling. Additionally, the dust can contain conductive elements that can cause partial short circuits in a system. Other elements in dust and dirt can promote thecorrosion of electrical contacts and connectors, resulting in improper connections. Regularly blowing out any dust and debris from the cooling inlets and outlets will benefit that device in the long run. Modern dusters use either HFCs (hydrofluorocarbons such as difluoroethane) or carbon dioxide, neither of which is known to damage the ozone layer, can be used for cleaning.

For preventive maintenance procedures, smoke may require more frequent cleanings, depending on the amount of dirt and dust in the environment.

Industrial Factors:

OSHA and IAEA keep standard operating procedures for the hazardous electrical equipment in industrial premises. These operating practices take the form of prescribed procedures for handling, operating, and maintaining specific pieces of equipment in a safe manner. For safety procedures, equipment should incorporate:

- Warning signs that list the electrical hazards present in the area and the emergency shut-down procedures.
- Insulating mats for personnel to stand on when working with electrical equipment.
- Barricades or barriers to prevent personnel access to areas of potential electrical hazard.
- Fire extinguishers to control electrically initiated fires.
- Test equipment to monitor the condition of electrical equipment. This can include insulation testers, a voltmeter that checks the condition of the insulation on wires and cables, and ground-circuit testers that indicate the presence of open circuits, shorted wires, or faulty grounds. For digital equipment, an appropriate oscilloscope and frequency counter are required.
- Ground fault interrupters (GFI) that protect a person who comes into contact with live components. A GFI detects the flow of current that results from the accidental contact and kills the circuit. A GFI can operate in a time as short as 25 milliseconds, thus minimizing the total current the person is exposed to.
- An insulated stick or pole for moving a potential victim away from a source of electricity.
- Some industrial building should retain devices for measuring noise, electro magnetic interference (EMI), vibration, corrosion, and power quality. Power quality is related to the total harmonics, pulses, power factors, grounding, imbalanced electrical loads, voltage surges and sags, resonance, etc.
- Residual current devices (RCDs) that provide protection against electrical currents flowing to earth through a person's body and present a shock risk, or flow through wiring or electrical appliances present a fire risk

Chapter 4 – Magnetic Fields and Human Behaviour

The discovery of electricity and electromagnetism, as well as their effects on humans and animals, show a variety of features. Electricity and electromagnetism are forces of nature that can be of benefit or drawback, depending on their application and purpose. In medicine, electrotherapy can assist the healing process if correctly applied. In nuclear fusion reaction, a magnetic field could be a disastrous to plants, animals and humans.

In addition, electromagnetic fields do possess an inner spiritual quality that must not be allowed to take hold of the human being completely. And exactly as electrical and electromagnetic phenomena display their effectiveness outwardly in physical matter, so they influence inwardly the moral constitution of human beings in the absence of their having taken protective measures against them, ('Electromagnetic Radiation and the Human Being', in: Trans Intellegence Internationale (Issue 3/4 1999), pp. 35-37).

For example, highly refined electromagnetic fields can accelerate the body's natural anti-inflammatory response, thereby aiding wounds to heal faster. This is showing some promising results. One animal study used a prospective, randomized, double-blind, placebo-controlled design to evaluate the effect of a specific noninvasive radiofrequency-pulsed electromagnetic field signal on tendon tensile strength at 21 days after transection in a rat model. This study found an increase in tensile strength of up to 69 percent (136.4 + 31.6 kg/cm2) at the repair site of the rat's Achilles' tendon 3 weeks after transection and repair, compared with the value (80.6 + 16.6 kg/cm2) in nonstimulated control animals. Although electroceuticals are promising, additional research is needed to recommend them for pressure ulcer treatment (http://www.ahrq.gov/qual/nurseshdbk/nurseshdbk.pdf).

Inwardly, our consciousness is a result of the dense electromagnetic field that surrounds the earth. The magnetic field of the earth is the product of the chemical force that is produced during the transformation of light into nothing inside matter. Light, as defined by Rudolf Steiner, decays and the decaying light is electricity. What we know as electricity is light that is being destroyed in matter. This is how the magnetic field of the earth is formed.

Since the light is destroyed in our bodies, it creates a magnetic field in ourselves that affects our consciousness which is a sense of one's personal or collective identity, including the attitudes, beliefs, and sensitivities held by or considered to be a characteristic of an individual or group: Love of freedom runs deep in the national consciousness.

It is well known that very high frequency such as Gamma, X-ray, or ultraviolet in even small quantities and for short durations have consequences that are detrimental to health. However, far less attention is paid to the question of which frequencies influence a human being's thinking, feeling and willing.

No one feels the interaction between such high frequencies entering the body and the magnetic field that envelops the earth. Also, no one can sense the interaction between such high frequencies and those charges emitted from the interior of the earth that constantly carry particles with either a positive or a negative electrical charge, ions, and other related forces. There is an assumption of a rotating core of liquid iron in the interior of the earth that envelops the planet in a magnetic dipole field, similar to that of a bar magnet. The magnetosphere varies from location to location (dense at the two poles), month to month, and from sea level to mountains. It also depends on the moon cycles. Accordingly, every point on earth is exposed to a constantly changing magnetic field with unchanging magnetic declinations towards the two magnetic poles, while its daily variations control the biological rhythm. To conclude, there is electrodynamic interaction between the surface of the earth and the human body. Such an interaction varies along side the latitude and altitude. It is measured in micropulsation in the 10 Hz zone.

It was found that pulse modulation; including small frequencies, could induce changes in walking and sleep, and in the brain physiology. It was also found that hemoglobin, the blood protein that carries oxygen, is weakly diamagnitic and is repulsed by magnetic fields. It means that the blood becomes more negative.

Magnet therapy is now using products including magnetic bracelets and jewelry, magnetic straps for wrists, ankles, knees, and the back; shoe insoles; mattresses; magnetic blankets (blankets with magnets woven into the material), magnetic creams, magnetic supplements, plasters/patches and water that has been "magnetized". Chinese practitioners claim that the magnets can restore the body's "electromagnetic energy balance" and can improve blood flow in underlying tissues.

In magnetic fields of high frequency (microwave), mobile phones and data communication networks (including digital wireless systems) propagate electromagnetic radiation in the microwave range. The WHO (World Health Organization) has classified mobile phone radiation on the IARC (International Agency for Research on Cancer) scale as Group 2B- possibly carcinogen. That means that there "could be some risk" of carcinogenicity, See my book "Chemistry, Biology and Cancer: the Bond."

The following are some of the biological changes caused by electromagnetic fields (radiation), as studied and reported in various researches.

Article Source: http://EzineArticles.com/6817917

- Protein changes in skin- protein can be increased or decreased
- Excited brain cells- some cells of the brain cortex (adjacent to the side of phone use) can be exited and others inhibited
- DNA damage- DNA can be broken at the phosphate back bone. The damage would therefore be passed on to future cells which could predispose them to becoming cancerous.
- Brain cell damage-a study on rat's brains showed damage to the neurons (brain cells) in various brain parts, including the cortex, hippocampus and basal ganglia.
- Aggressive growth in leukemia cells- researchers at the National Research Council in Bologna, Italy found that Leukemia cells exposed to cell phone frequencies (900mH) for 48 hours replicated more aggressively.
- Increased blood pressure- researchers in Germany found that one-time use of a cell phone for 35 minutes could cause an increase in resting blood pressure of between 5 and 10mm Hg. Lancet June 20, 1998
- Brain consciousness communication to the outside world - It has been demonstrated that there is a relationship between electromagnetic fields, information, the phenomenology of consciousness and the meaning of free will (http://www.ingentaconnect.com/content/imp/jcs/2002/00000009/00000 008/1298).
- The conscious electromagnetic field theory (cemi field theory) proposes that consciousness is a manifestation of the brain's electromagnetic (em) field. The key feature of the brain's em field is that it is capable of integrating vast quantities of information into a single physical system and it thereby accounts for the binding of consciousness.
- Heart pace maker- a study on heart resonance of a coherent heart (pace maker) showed that the coherent heart of one person can have a scientifically demonstrable effect on the brain waves of another person (http://lawyertopeacemaker.com/heartmath.html).

4.1 Earth's Magnetic Field

The Earth has a magnetic field that originates from its south magnetic pole, extends out into space and comes back to its north magnetic pole. The north and south magnetic poles are near the north and south celestial poles about which the earth rotates (we stand vertically on the horizontal line connecting the geographical north and south poles), Figure (4.1). This is not the case of some other planets. The magnetic field around the earth is similar to one that would be formed if a bar magnet extended through the center of the earth with the tips at the north and south magnetic poles. According to accepted theory, Earth's magnetic field arises from electrical currents flowing in the molten iron core of the planet.

The earth's magnetic field extends out into space forming the magnetosphere. As the solar wind expands out from the sun, it encounters the magnetic field of Earth. On the sunward side of Earth, the solar wind compresses the magnetic field in toward the Earth, increasing the magnetic field strength in the compressed areas. On the opposite side of earth, the solar wind acts to stretch out the magnetic field thus giving it a teardrop shape.

The universe seems to be designed specifically for life. Physical constants such as the strength of gravity, the mass of the proton, and the charge of electrons appear so finely tuned that the smallest variation would prohibit life in the cosmos. However, the changes of such physical constants on the surface of the earth are negligible.

The gravitational field, which defines the measure and curvature of spacetime, is determined by the sum of all forms of energy and momentum. It was found that the gravity is not just the mass-energy density but actually the energy density plus three times the pressure as per the following formula:

$$G = V_e + 3V_p$$

Where G is the gravity, V_e is the vacuum energy and V_p is the vacuum pressure.

So, a person on the surface of the earth is exposed to many energy forces like gravity, cosmic radiation, earth magnetic field, and human magnetic field. The most important energies are the cosmic and the earth magnetic fields. The energy of gravity that can be added depends on the environmental conditions and the location on earth.

Figure (4.1) illustrates the two main energies affecting the human body in three different locations.

Figure (4.1): Types of energies affecting human body

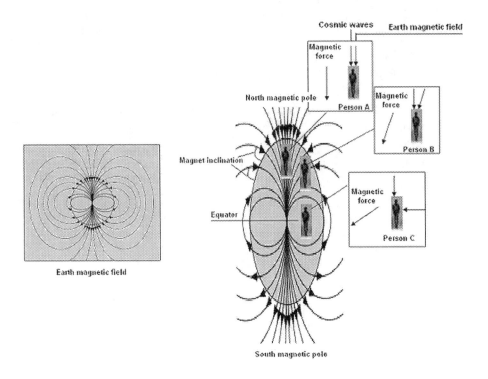

The magnetic field of earth is involved in a perturbation or morphing. This is taking place interdimensionally, and the net effect of all surrounding waves is incremental surges in amplitude or strength of the magnetic field. This morphing of earth's magnetic field is like a wave that rises and falls very quickly, and the oscillations are minute, but distinct. This is having a very strange effect upon human consciousness, specifically biological experiences. Many people are experiencing an increase of exhaustion and weariness both of which are caused by movements of energies from deep space as they pass through the galaxy and the solar system (http://tomkenyon.com/earths-magnetic-field).

Another set of symptoms directly related to the perturbations of your magnetic field are distinct changes in how you hold short term memory. Memory is a function of the magnetic fields of your own nervous system and those of the total forces coming from the inside of the earth. Cosmic rays, or peripheral waves reach the north pole from the south pole. Your brain processes information through the minute gravitational and non-gravitational fields of its own structure and it is affected directly by fluctuations in the earth's magnetic field and other fields. Therefore, the relationship between Earth, cosmic magnetic fields, and cellular function of the body is a key component to the understanding of the conscious evolution (http://www.angelfire.com/ms/MelchizedekAngel/magnetic.html).

Measurements of magnetic fields indicate a decline in intensity from 8.5 x 10 to 8 x 10 in the 25 units over 130 years. In other words, a decline in the magnetic field of about 5% every 100 years has happened. This decline is due to the rapid expansion of the universe.

The National Bureau of Standards in Boulder, Colorado reset the Cesium-atomic clocks to reflect "lost time" in the day-- days are becoming longer than the clocks can account for. A "cesium atomic clock" is the most accurate clock that is used as a reference; using the frequency of the microwave spectral line emitted by atoms of the metallic element cesium, in particular its isotope of atomic weight 133 ("Cs-133"). The integral of frequency is time, so this frequency (f), is 9,192,631,770 hertz (Hz = cycles/second). The time is proportional to 1/f. The frequency of the microwave on earth depends on the temperature of the atmosphere and the distance from the sun. The expansion of the universe causes the earth to move away from the sun, and results in lower temperatures and less frequency. Thus, our recent days are longer than the past days.

Imagine what will happen to signals from an object to our retina when the day becomes longer and the frequency of visible light becomes slower (less frequency means slower)? Signals travel along the nerves as waves of electrically charged ions. When they reach the nerve terminus, the signal jumps to the next nerve via axons carrying chemical neurotransmitters.

The electrical ions (signals) definitely will match the new frequency of the new day and are processed in our brain. The processed data will have an effect on our consciousness and self-awareness.

As pointed out earlier, each zone of the earth has its own magnetic force and cosmic waves. The magnetic waves in the two pole zones are denser than those in the equator zone, therefore the frequency in there would be higher. A new study suggested that the speed of transmission of electrical ions would be faster in the brains of people living on both north and south poles. However, the signals would slower on the equator and become more liked; i.e. more consolidated.

4.2 Magnetic Medicines

Preliminary scientific studies of magnets for pain have produced mixed results. Overall, there is no convincing scientific evidence to support claims that magnets can relieve pain of any type. Some studies, including a recent

National Institutes of Health (NIH) clinical trial for back pain, suggest the possibility of a small benefit from using magnets for pain.

In November 1917, in St. Galen, Rudolf Steiner spoke about the important future task of establishing a new geographic medicine, based on the interaction of the electromagnetism of Earth and humans.

Across the world magnetic bracelets or wraps and other similar items are becoming popular. One company uses testimonials from sports figures who claim, "Tectonic® magnets are worn by football players, pro tour golfers, tennis players and active people world wide." Magnets in products like magnetic patches and disks, shoe insoles, bracelets, and mattress pads are used for pain in the feet, wrist, back, and other parts of the body. Magnets are generally considered safe when applied to the skin. They may not be safe for people, who use medical devices like pacemakers or defibrillators, as magnets may interfere with the device. Magnets marketed for pain usually claim strengths of 300 to 5,000 G—many times stronger than the earth's magnetic field (about 0.5 G) and much weaker than the magnets used for MRI machines (approximately 15,000 G or higher).

The effect of magnets is based on repeated motion and pulsating energies that cause beneficial chemical changes in the body. However, the majority of rigorous studies have found no effect on pain. More research on magnets for pain is needed before reaching any firm conclusion.

Scientific researchers and magnet manufacturers have proposed that magnets might work by:

- Changing how nerve cells function and blocking pain signals to the brain
- Restoring the balance between cell death and growth
- Increasing the flow of blood and the delivery of oxygen and nutrients to tissues
- Increasing the temperature of the area on the body being treated.

The side effects and risks of magnets may not be safe for some people who use a pacemaker, defibrillator, or insulin pump, because magnets may interfere with the functioning of the medical device. Magnets should not be used with people who have wounds that have not yet healed. Otherwise, magnets are generally considered safe when applied to the skin. Reports of side effects or complications have been rare.

The National Center for Complementary and Alternative Medicine (NCCAM) is carrying out research using magnets on the following projects:

- Carpal tunnel syndrome (pressed or squeezed wrist)
- Fibromyalgia (debilitating fatigue, sleep disturbance, and joint stiffness)
- Knee osteoarthritis
- Low-back pain (lumber)
- Networks of blood vessels involved in healing.

4.3 Meridian Nerves

Nerves work like electrical wires that carry information from the brain to the rest of the body and vice-versa. They are distributed throughout the entire body. They include:

- Motor (efferent) nerves which carry data from the brain out to the various organs of the body. For example, commands (electrical current) are sent to the heart, causing the chambers to squeeze and release in a steady, rhythmic sequence that draws blood into the heart and pushes it out.
- Sensory (afferent) nerves which carry signals from the periphery of the body to the brain or spinal cord for processing, including information about pain, touch, taste, temperature, or other sensations.

Over 5,000 years ago, the ancient Chinese discovered a subtle energy in the body that can't be seen, felt or found with the senses. Matter and energy are two different manifestations of the same primary energetic substance of which everything in the universe is composed, including our physical and subtle bodies. For example, the black hole in the universe is matter, whereas the black energy is a subtle energy. As per the relativity theory, matter exists when it moves at frequency below the speed of light, but it disappears if it moves at speeds exceeding light velocity. In this case, it is known as subtle matter. As per the ancient Chinese description, subtle matter is as real as dense matter; its vibratory rate is simply faster. It is believed that two opposite ends of the spectrum--yin, the energy of earth and yang, and the energy of heaven--combined with humans to create this vital force.

The Chinese discovered and identified twelve acupuncture meridians along which this energy travels in the human body. Acupuncture meridians are like copper traces on an electronic circuit board, running throughout the body. They were named by the life function associated with them. To the majority of Western scientists, acupuncture meridians seem like imaginary structures because there are no published anatomical studies of the meridians in orthodox medical journals to substantiate their existence. Meridians are the pathways of the positive and negative energy power, which carries on some of the communication between the various parts of human beings, http://tuberose.com/meridians.html

Meridians connect specific teeth, organs, and tissues. In fact, they connect everything in the body. These have been measured and mapped by modern technological methods, electronically, thermatically and radioactively. Normal skin resistance over a healthy point is 100,000 Ohms. With practice and awareness they can be felt. Through these meridians passes an invisible nutritive energy known to the Chinese as Ch'i. The Ch'i energy enters the body through specific acupuncture points and flows to deeper organ structures, bringing life-giving nourishment of a subtle energetic nature. Acupuncture points have unique electrical characteristics, which distinguish them from surrounding skin. These acupuncture points exist along the meridians. These points are electro-magnetic in character and consist of small palpable spots, which can be located by hand, with micro-electrical voltage meters and with muscle testing, when they are abnormally functioning.

These 500 points, mapped and used for centuries to optimize human performance, are connections between the positive and negative meridians and functions of the body, including internal organs and muscles. These points are useful not only in treatment but also in the diagnosis of disease states. Subtle magnetic Ch'i currents flowing through the acupuncture meridians are not electrical in nature, but they are able to induce secondary electrical fields that create measurable changes at the physical cellular level through the induction of secondary electrical fields. The meridian nodes (Yin and Yang) are shown in Figure (4.2) (http://tuberose.com/meridians.html).

According to western medicine, our internal organs and our external bodies are connected by nerves. According to the traditional Chinese medicine, our internal organs and our external body are connected by the meridians. Thus, there are dual pathways (meridians and nerves) connecting the surface of your body to your internal organs.

Pressure applied to certain points along meridians in the body allows the body to balance and heal itself. Meridians are pathways recognized by traditional Chinese medicine that carry energy throughout the body. Each is named for an organ and the points along it relate to the functions of that organ in ways that may not seem familiar in Western medicine. To stimulate an acupressure point, use the tip or second knuckle of your index finger and press on the point for at least 30 seconds. The meridians run vertically through the body, beginning or ending in either the hands or feet. Reflexology affects the meridians, as many of the reflexes are actually on meridian lines. These energetic pathways connect with the systems of the body. An obstruction along a meridian may disrupt the function of organs found on that channel. Six meridians run on the legs — three yin and three yang. Six more run on the arms with the same divisions.

Figure (4.2): Meridian nodes on the human body

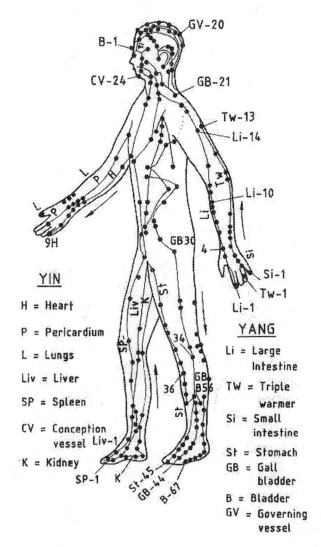

4.4 Electromagnetic Interference

Electromagnetic interference (EMI) is a disturbance that interrupts, obstructs, or degrades watching television, listening to the radio or talking on your cordless telephone. It affects an electrical circuit due to either electromagnetic induction or electromagnetic radiation emitted from an external source. These effects can range from a simple degradation of data to a total loss of data. EMI can be caused by an artificial or natural high frequency or a radio frequency that changes electrical circuits. Artificial causes can be arc furnaces, short circuits, or switching high voltage circuit breakers. Natural causes are lightning, sun storms, or the Northern Lights. In electronic warfare,

EMI can be used to jam signals on radios, rocket launchers, or spurious emissions used for the protection of safety services from unwanted emissions.

In the recent 20 years, however, there has been growing awareness that electromagnetic interferences were affecting the human body. Most notable was the concern that cellular phone usage and high power lines caused cancers and diseases. In addition, some experts claim that exposure to low-frequency electromagnetic fields (common in household appliances) may disrupt normal cycles and the production of the hormones melatonin and serotonin. In Europe, loads of non-linear type (inductive or capacitive, but not resistive) must have filters built into it at the factory. As a result, American computer manufacturers must put an inexpensive RF (radio frequency) filter in the computers they export to Europe.

4.4.1 Electromagnetic Interferences Affect the Nervous System

The frequency of the electromagnetic interference (EMI) is a very important factor that determines its interaction with the human body. Cellular phone radiations are at higher frequencies than transmission line radiations and they can affect human nerve tissues. High frequencies tend to penetrate less and heat human tissue, just like microwave oven frequencies that heat the upper layers of foods and leave the lower layer colder. It has been hypothesized that high frequency electromagnetic energies might interfere and resonate with DNA and only other cellular apparatus which might trigger cancerous changes.

The US government tested high power microwave weapons that had the ability to disrupt animal and human behavior; which was attributed to heating the brain with microwaves, Figure (4.3).

Figure (4.3): The brain and abdomen can be heated by electromagnetic interference

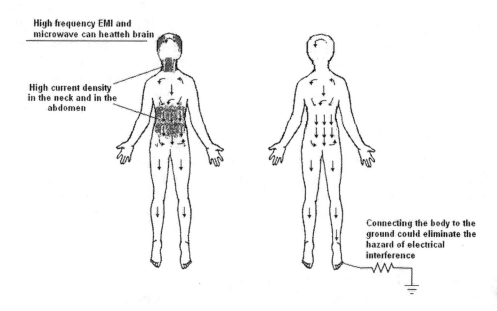

The crudest of these armaments would be a sort of electromagnetic flamethrower with a greater range than chemical types. Dogs were cooked to death in experiments at the Naval Medical Research Institute as long ago as 1955. High-power transmitters using short UHF wavelengths can severely burn exposed skin in seconds.

Inside a typical house the low frequency electric field is usually 100 times higher than the one outside. Outside household's earth, buildings, trees and other materials interposed with the electromagnetic sources attenuate and shield from the electric field, but not the magnetic field. Even the buried of power lines radiate magnetic fields (http://www.biotele.com/EMI.htm).

Chapter 5- Cosmic Electromagnetic Radiation and the Human Body

Electromagnetic radiation (light waves that are fluctuations of electric and magnetic fields in space) can be described in terms of a stream of photons, each traveling in a wave-like pattern, moving at the speed of light (299 792 458 m/s or 186,282 miles/second) and carrying some amount of energy.

5.1 The Spectrum of Electromagnetic Radiation

The spectrum of the electromagnetic radiation is usually divided into seven parts: radio waves, microwaves, infrared, visible, ultraviolet, X-rays and gamma-rays. Cosmic rays are spectrums of higher frequencies than gamma-rays. Table ( ) shows the characteristics of the above spectrums.

Spectrum	Frequency Range (Hz)	Wave Length	Atomic Transition	Characteristics
Cosmic rays	Above 10^{24}	$>10^{-12}$ m	Hydrogen	High energy range between 10^{13}-10^{16} eV
Gamma rays	10^{20} - 10^{24}	$<10^{-12}$ m	by sub-atomic particle interactions such as electron-positron annihilation, neutral pion decay, radio active decay, fusion and fission in astrophysical process	Unlike optical light and X-rays, gamma rays cannot be captured and reflected in mirrors. The high-energy photons would pass right through such a device
X-rays	10^{17} - 10^{20}	1nm – 1pm	Inner electrons	X-rays are electrically neutral. They have neither a positive nor a negative

				charge. They cannot be accelerated or made to change direction by a magnet or electrical field.
Ultraviolet UV-A (320-420nm), UV-B (280-320 nm), and UV-C (less than 280 nm)	10^{15} -10^{17}	400nm – 1nm	Outer electrons	The cumulative exposure of UV-B radiation may cause sunburn, cataracts, suppressed immune systems, and premature aging including; wrinkles and skin discolorations as well as skin cancer.
Visible	4 - 7.5 x 10^{14}	750nm – 400nm	Outer electrons	Visible light can cause chemical changes in some materials
Infrared	10^{13} - 10^{14}	25µm – 2.5 µm	Outer electrons or molecular vibration	Infrared is used as a powerful tool for determining the internal structure of molecules and for

				identifying the amounts of known species in a given sample
Microwave	$3 \times 10^{11} - 10^{13}$	1mm – 25 µm	Molecular rotations	One of the basic characteristics of microwave energy is that it is reflected by metal
Radio waves	$<3 \times 10^{11}$	>1mm	Molecular spin flip by magnetic fileld	Radio waves move in free space and over the surface of the Earth

Electromagnetic radiation from the universe (gamma and X-ray) could reach the surface of the earth together with the visible spectrum, radio frequency, and some ultraviolet wavelengths. Astronomers and astrophysicists use balloons, rocket flights or electromagnetic detectors on orbiting satellites to observe other spectrums. Figure (5.1) illustrates the spectrum wavelength versus altitude.

Figure (5.1): The spectrum wavelength and altitude

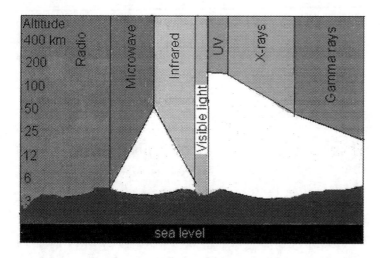

In the past three decades, observational astronomy has expanded from the relatively narrow wavelength band of visible light, which is one octave in width (octave is doubling the frequency), to the entire electromagnetic spectrum. Today, more than sixty octaves between the long-wave radio band and the range of high-energetic gamma-ray radiation are used. The mainspring of this development was the awareness that different spectral ranges allow different and complementary insights into cosmic events.

The electromagnetic force binds electrons to atomic nuclei (clusters of protons and neutrons) to form atoms. The gravitational force acts between massive objects. Although it plays no role at the microscopic level (between atoms), it is the dominant force in our everyday life and throughout the universe.

Stream of protons and electrons flow from the sun corona towards space at a speed range between 300 and 1000 km per second are also magnetic forces. The streams come from holes in the sun's corona, and push the gas of the comets' tails away from the sun, causing geomagnetic disturbances on the surface of the earth. Geomagnetic storms are other phenomena that would influence the environment, biological reaction, telecommunication and power grid lines.

Gamma ray bursts are the name given to flashes of Gamma rays emanating at random from distant galaxies. They are the most radiant electromagnetic fields in the universe after the occurrence of the Big Bang. Their duration is typically a few seconds and they can release more energy in 10 seconds than the sun will emit in its entire 10 billion-year lifetime! So far, it appears that all of the bursts we have observed have come from outside the Milky Way Galaxy. By solving the mystery of Gamma ray bursts, scientists hope to gain more knowledge of the origins of the universe and the rate at which the universe is expanding

In a positive electrostatic environment, the amount of positive ions (cations) in the air is so high that the positive ions in our body also increase. As a result, our body will fail to undergo metabolism, our cells become weak, we will be easily affected by illnesses and the aging process will accelerate. As you can see in our current modern lifestyle, our body is exposed to positive ions generated by cell phones, ultraviolet rays, acid rain, chemicals, electronic equipment, and computers.
Conversely, grasslands, forests, and tress, are high in negative ions, especially in waterfall areas, which is why these places are known to be the best environment for your health.

For static magnetic fields, acute effects are only likely to occur when there is movement in the field. This could be the motion of a person or internal body movements like blood flow or heart beat. A person moving within a field above 2 T can experience sensations of vertigo and nausea, and sometimes a metallic taste in the mouth and perceptions of light flashes occur. Although only temporary, such effects may have a safety impact for workers executing delicate procedures (such as surgeons performing operations within MRI units).

Static magnetic fields exert forces on moving charges in the blood, such as ions, generating electrical fields and currents around the heart and major blood vessels that can slightly impede the flow of blood. Possible effects range from minor changes in heartbeat to an increase in the risk of abnormal heart rhythms (arrhythmia) that might be life-threatening (such as ventricular fibrillation). However, these types of acute effects are only likely within fields in excess of 8 T.

It is not possible to determine whether there are any long-term health consequences even from exposure in the millitesla range because there are no well-conducted epidemiological or long-term animal studies to date.

Certain animal species appear to be able to use the earth's magnetic fields for orientation. Such effects are not present in humans, therefore this has no implications on human health. There is good evidence that exposure to static magnetic fields can induce electric charges around the heart and major blood vessels. This has been observed in smaller animals when fields exceed about 1000 mT (and in larger animals 100 mT), but possible health consequences are unclear. In pigs, several hours of exposure to very high fields (up to 8000 mTT) did not result in any such effects. However, in rabbits, exposures to much lower fields have been reported to have a slight influence on the heart rhythm and the circulation of the blood, although the evidence for this is not strong.

Few studies have looked at the possible effect of static magnetic fields on blood cells, on the endocrine system, or on reproduction and development. Most of the results are not consistent and have not been replicated, so further studies are needed in order to assess the health risk.

In general, so few animal studies have been carried out with regards to harmful effects on genetic material and cancer, that it is not possible to draw any firm conclusions from them. Cancer researchers are concerned with power lines that can cause effects on cancer at 1 mG (milligauss) if people are exposed to them more than 6 hours per day. Power lines of 15 kV and above are the ones of concern. Leukemia cancer studies link low levels of EMF fields of 2 -4 mG to the development of the cancer. A study (Ahlbom and others) reported that at 2 mG and above, exposed children were 2.7 times as likely to develop cancer as unexposed children, and those exposed to 3 mG and higher were 3.8 times as likely.

5.2 Safety Guidelines for EMF (Electro Magnetic Exposure)

The International Commission of Non-Ionizing Radiation Protection (ICNIRP) published safety guidelines for EMF (Electro Magnetic Field) exposure, Table (3).

Table (3): Electric fields and magnetic fields as classified by ICNIRP

Exposure (60 Hz)	Electric field	Magnetic field
Occupational	8.3 kV/m (kilovolt per meter)	4.2 G (Gauss)
General public	4.2 kV/m	0.833 G

Note 1: Computer monitors and VDTs (video display terminals) must not be more than 2 mG.

Note 2: Microwaves should not be more than 0.1 mW/cm^2 or 1mG at 1ft distance.

Note 3: There are many Gauss meters available in the market for measuring the Gauss level, ranging from $100 to $ 800 and more depending on the resolution and accuracy of the meter.

Here are some precautions and recommendations for avoiding high level of radiations:

a. Measure your home, work and school places with a Gauss meter. Don't let your children play or gather near microwave towers, power lines of 15 kV and above, distribution transformers, power substations or power generation plants, and cogeneration plants, and heavy industrial plants such as (steel mills) that have arc furnaces and welding and melting machineries,

b. Avoid areas and places where the magnetic field is above 1 mG. Measure the Gauss level from appliances and electrical machines both when they are turned off and when they are operating. For example, dimmer switches or TVs have radiation even when they are off. They may have no magnetic fields, but they may have electrical fields (Electric and magnetic fields are so closely related that we sometimes describe them as a single entity, "electromagnetism", in most cases they are perpendicular to each other).

c. Don't sit too close to your TV set. Distance yourself at least 6 feet (about 2 meters) away. Use a Gauss meter to locate where it is safe to sit.

d. Don't sit too close to your computer. Computer monitors or CPUs vary greatly in the strength of their EMFs due to the size of the screens and methods of conversion and switching, so you should check yours with a Gauss meter. Don't stand for long periods close to your microwave oven. Move all electrical appliances with moving or revolving parts (incandescent lamps are ok, however, fluorescent lamps and other discharged lamps have ballasts which release magnetic radiation due to its impedance resistance nature) at least 6 feet from your bed. Eliminate wires running under your bed. Eliminate dimmers and 3-way switches. Detect low levels of radiation by using a radio instrument switched to MW (medium wave) frequency which is disturbed by low radiation.

e. Don't use glasses with metallic frames. They should be made from plastic with no metallic wires in them, otherwise they could serve as antenna to receive and send radio and cellular phone waves into your brain.

f. Avoid exposure to Beta, X-ray, Gamma ray, and neutron rays as all of them penetrate the human body, and could affect the cells and the genetic components of DNA and RNA, Figure (5.2).

Figure (5.2): Exposure of the human body to cosmic waves

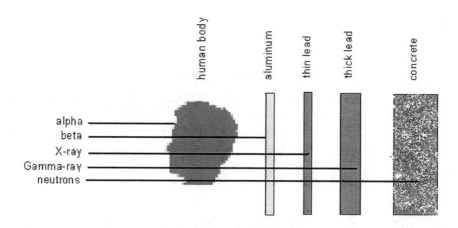

Studies on health effects due to exposure in the workplace almost exclusively focused on workers exposed to moderate static magnetic fields generated by equipment using large DC currents (aluminum smelters). Among such workers, increased risks of various cancers have been reported, but results are not consistent across studies. These workers are exposed to a variety of other potential hazards, which makes the exact cause of any observed effects unclear. The data available so far is inadequate for a health evaluation.

5.3 Perception of Light Flashes

Other short term effects of static magnetic fields have been reported, but since the experiments have not been repeated by other researchers in order to test the validity of the results it is difficult to draw any firm conclusion. Furthermore, there is not enough evidence to reach any conclusion on long term effects like cancer. Immunodeficiency can be caused by medicinal drugs. For example, medicines given to patients with transplant organs that suppress the immunity could cause immunodeficiency. Radiation and chemotherapy treatment for cancer attacks both cancerous cells and healthy cells.

5.4 Cell Division

Some cells divide rapidly. For example, beans take about 19 hrs to complete the cycle division, whereas red blood cells divide at a rate of 2.5 million per second. Cancer cells divide rapidly; the daughter cells divide before they have reached maturity. Electrocharge, pH, temperature, and some drugs (enzymes) may affect the rate of division. When cells stop dividing, they stop at a point late in the stage G1. The stage S is the stage when the DNA is replicated for the next division, and the chromosomes become double stranded. The cell has then entered the G2 stage and proceeds in to cell division. Cells will not divide again and stop in the G1. The process of division passes into the following steps:

5.4.1 Prophase

Prophase is the first step of mitosis preparation where the cell is about to divide, and the chromosomes become visible and start to condense into double stranded chromosomes, Figure (5.3). Chromatin/DNA do not replicate in this step. Gradually, a spindle composed of protein fibers together with kinetochores forms and extends nearly the length of the cell, expanded in its centre (or the equator of the cell) like a base ball. The chromosomes each consist of two chromotids attached to each other by a spindle fiber at the centromere.

Figure (5.3): Double stranded chromosome divides into two identical sisters.

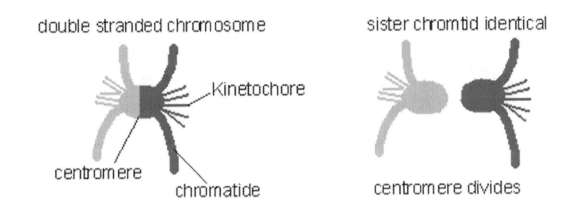

5.4.2 Metaphase

When the centomere and kinetochore arrive at the centre of the cell (equator), the metaphase begins. The mitosis process ends when the centromere divides so that each of the chromatids becomes a single stranded chromosome, Figure (5.3).

5.4.3 Anaphase

During anaphase, each sister of a single-stranded chromosome moves towards one pole of the cell (one sister to one pole, and the other to the opposite pole). Thus, anaphase is the stage when sisters of chromosome migrate to the two poles of the cell, Figure (5.4).

Figure (5.4): Movement of daughter chromosomes from equator to poles during anaphase

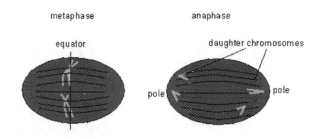

5.4.4 Telophase

In telophase, the nuclear envelope (which had disappeared during prophase) reforms, and chromosomes uncoil into the chromatin form again. There are now two cells instead of one cell, but they are of smaller sizes of the same genetic heredity. The small cells will develop into mature ones.

5.4.5 Cytokinesis

Cytokinesis is the last stage of cell division, where the daughter cells split apart. The mitosis is the division of the nucleus, and the cytokinesis is the division of the cytoplasm.

5.5 Ionic and Isotopic Radiation

In medical diagnosis and research, isotopes of iodine are used in the diagnosis of thyroid functions, in the treatment of hyperthyroidism, and in the cathodic - ionic blood circulation action and reaction of substances for research.

An isotope is an element whose nucleus contains a specific number of neutrons, in addition to the number of protons that uniquely defines the element. It means that an atom of the same element can have a different number of neutrons and a fixed number of protons. An atom with a different number of neutrons is called isotope. Hydrogen has no neutrons at all. However there are some hydrogen isotopes; deuterium (hydrogen with one neutron), and tritium with two neutrons, Figure (5.5).

Figure (5.5) Isotopes of hydrogen

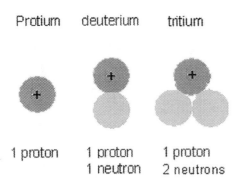

Protium   deuterium   tritium

1 proton   1 proton   1 proton
           1 neutron   2 neutrons

Isotopes are written in the form ^AB_C, where B is the symbol of the element, C is the atomic number, and A is the number of neutrons and protons combined, called the mass number. For example, ordinary Hydrogen has the formula of $^1H_1$, deuterium $^2H_1$, and tritium is $^3H_1$. Every chemical element has more than one isotope, and one isotope is more abundant in nature than any of the others. Multiple isotopes of one element can be found mixed.

Certain isotopes of elements are unstable, radiating energy or radioactive waves. These are called radioisotopes. Carbon - 14 ($^{14}C_6$) is a radioisotope. Certain isotopes are more radioactive than others. There are many carbon isotopes ranging from 8C to 22C. All of them have halftimes (will be discussed soon) ranging from several seconds to microseconds, except the 14C which has 5730 years of halftime. Radioisotopes have an unstable nucleus that emits rays of types alpha, beta, or gamma (will be explained) until the element reaches stability point.

The stability point in this case is a nonradioactive isotope of another element. For example, radium - 226 ($^{226}Ra_{88}$) decays finally to lead - 207 ($^{207}Pb_{82}$). Precise measurement shows that some elements still contain traces of radioisotopes even with light elements such as hydrogen. Isotopes of phosphorous are used in medical therapy to kill cancerous cells, such as leukemia (increase in white blood cells) and polycythemia (increase in red cells).

Radioisotopes release energy (decay) by spitting out energy in the form of particles or electromagnetic waves.

Let's take some examples of ions:

1- Carbonate or carbon trioxide CO_3^{-2}

Carbonate works as a modifier of pH in the blood. When pH is too low it absorbs hydrogen ions and causes the CO_3 to shift left, i.e.

less negative, and accordingly the blood will get rid of (exhale) CO_2, and this will cause decreasing level of H^+. When the pH is too high (alkaline), the hydrogen is reduced, and more oxygen is inhaled. Thus the pH is balanced.

Resonance structure (will be discussed) can be used to depict the carbonate ion. The oxygen with a double bond has eight electrons in the outermost orbit, but the oxygen with a single bond has only seven, and therefore is negative.

2- Nitrate NO_3^-

Nitrate is toxic to humans because it oxidizes the iron in hemoglobin. The oxygen and nitrogen want to take the two and the three electrons from the two outermost orbits (14, 2) since they are very weak with respect to nitrogen and oxygen. The iron would then have double or triple cations, rendering it unable to carry oxygen, as the remaining electrons are stuck to their last orbit, and the oxygen can not pull any electrons from the iron. This condition is called methomoglbinemia.

Sometimes drinking water contains high level of nitrates. Water that comes from springs close to land, could contain high levels of nitrate. Shallow rivers and creeks could have adverse effects on different aquatic species.

A nitrogen atom needs only three electrons for its outermost orbit to be stable. Oxygen is more negative than nitrogen and therefore it pulls one electron from the nitrogen, making the nitrogen positive.

3- Nitrite NO_2^-

Nitrate in food (meat and fish) delays the development of botulinal toxins that could affect the central nervous system and cause muscular paralysis, and retards development of rancidity and off-odors and flavors.

Nitrate can easily get rid of one oxygen atom to replace with one atom of sodium, as repulsion between nitrogen and sodium is much less than that between oxygen and nitrogen. The product is called sodium nitrite, which is commonly used for curing meat products. Part of sodium nitrate (nitric oxide) combines with myoglobin (muscle protein), the pigment responsible for the natural red color associated with cured meat. Nitrate sodium can be fatal if the dose is above the range of 23 milligrams per kilogram of the body weight. It has been reported that people normally consume more nitrates from their vegetable intake than from the cured meat products they eat. Some reports suggested that nitrate and nitrite are related to some types of cancer diseases, however, there is no confirmable evidence in medical literatures on the carcinogenicity of nitrate or nitrite.

4- Nitronium ion NO^+

Nitronium is a strong oxidizing agent and may cause burns to the skin or eyes. It may be harmful if swallowed, inhaled or absorbed through the skin, and is very destructive to mucous membranes. In vitro, it can combine with other oxidizing substances such as BO_4 and the produced component can be destructive to human cells.

115pm

180°

5.5.1 Alpha, Beta, and Gamma decay

Radioisotopes are continually undergoing decay.

Alpha decay is common with elements of an atomic number greater than 83; from rubidium onwards. Alpha decay can be written in the following form:

$$^{A}B_C \longrightarrow {}^{A}B_C - {}^{4}He_2 \longrightarrow {}^{A-4}D_{C-2}$$

Where B is the parent isotope and D is the daughter isotope and is different element than element B. Helium (He) is called the alpha particle.
The new element D is now less in atomic number by 2, and consequently goes back two places in the periodic table. Can we convert thallium (81) to gold (79)?
Helium has two electrons in the first orbit, and they are strongly attached to the nucleus, which has two positive protons. The two electrons are very reluctant to leave the helium atom. Therefore, alpha is not very penetrating as it captures electrons immediately from the substance it hits. However, it is very damaging because the energy released, due to the capture of more electrons, can knock atoms off the substance.
Alpha rays are used in many applications. Since alpha decay travels a short distance (few centimeters or inches), it can be used to ionize small air (mainly CO_2) and allow a small current to flow. As an example of this is smoke detector. CO_2 will be ionized to CO_2^- and a current of several milliamps will go through the gap of the smoke detector, sounding the alarm.

Alpha decay is used in thermoelectric generators used for space probes and artificial heart pacemakers, because alpha decay is much more easily shielded against than other forms of radiations. Plutonium -238 (the product of plutonium 244 : $^{244}Pu_{94} - {}^{4}He_2 = {}^{238}Pu_{92}$) requires only one inch (2.54 centimeters) of lead to protect against other radiations.

Beta negative decay is the process of converting the neutron into a proton, and it follows the equation:

$$^{A}B_C \longrightarrow {}^{A}B_C - {}^{0}e_{-1}{}^{-1} \longrightarrow {}^{A}B_{C+1}$$

The tritium (H-3) can be converted into helium, Figure (5.6), by converting one neutron into a proton. So $^{3}H_1$ becomes $^{3}He_2$, and an unstable isotope is converted into a stable isotope of helium, or semi stable helium. For complete stability, the number of protons must be equal to the number of neutrons.

Figure (5.6) Beta negative decay turns tritium into helium.

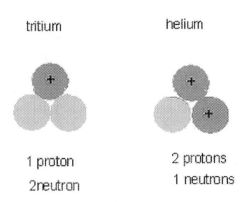

tritium

helium

1 proton

2neutron

2 protons

1 neutrons

Beta negative decay is fast and furious at the beginning and slow down over time. It is more penetrating than alpha because the particles are smaller.
In general, there are three forms of beta decay:

    a) Electron emission in which a neutron converts into a proton with the emission of an electron which is called a beta-minus particle. It takes the form:

$$A^0 \longrightarrow P^+ + e^-$$

Note that the total charge is the same on both sides of the formula. The electron e^- cannot exist inside the nucleus and therefore it is repelled and released to the outside of the nucleus.

When the electron is repelled, it can cause damage to the cell of a human body. So if beta decay is not properly shielded, the electron could oxidize the human cell and cause a mutation to it, and possibly end with a cancerous result. An example of this negative decay occurs in the iodine -131, which decays into xenon - 131 and one electron:

$$^{131}I_{53} \longrightarrow {}^{131}Xe_{54} + {}^0e_{-1}$$

Note that the alpha decay decreases the atomic number, whereas the beta decay increases the atomic number. In other words, alpha is a reducing agent, and the beta is oxidizing agent. This would relate to the Warburg hypothesis for cancer growth that will be dealt with in chapter 3.
The above equation shows that the number of protons has been increased, and if the number of protons has increased above a certain limit, the atom will not be stable. In this case, the atom attempts to be stable again by converting some protons back to neutrons with the emission of a positively - charged electrons.

b) An electron with a positive charge is called a positron:

$$P^+ \longrightarrow n^0 + e^+$$

$\longrightarrow$ positron

An example of this type of decay potassium - 39 which decays into Argon -39:

$$^{39}K_{19} \longrightarrow {}^{39}Ar_{18} + {}^{0}e_{+1}$$

- An electron capture or beta decay X-ray, in which an inner orbiting electron is attracted into an unstable nucleus and combines with a proton to form a neutron:

$$e^- + p^+ \longrightarrow n^0$$

In this process, there is no radiation emitted, but the cloud surrounding the nucleus is changed, and filled by an electron from an outer shell. The filling of the vacancy is associated with the emission of X-ray. That is called the beta decay of an X -ray type.

Gamma decay involves the emission of energy from an unstable nucleus in the form of electromagnetic or photon radiation. It follows the following formula:

$$^{A}B_C \longrightarrow {}^{A}B_C + {}^{0}\gamma_0$$

In gamma decay, a nucleus changes from a higher energy state to a lower energy state through the emission of electromagnetic radiation or photons. The number of protons (and neutrons) in the nucleus does not change in this process, so the original and the output atoms are the same chemical element, Figure (5.7).

Figure (5.7): Gamma change in energy

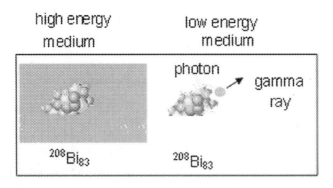

In alpha and beta decay, the atomic number of the nucleus changes, but in gamma decay the atomic number stays the same. In gamma decay, the change is only in the energy state of the nucleus, which will be changed to a lower state by emitting photons. The photons produced in this decay are consequently known as gamma rays and have a wavelength with an order of magnitude of about $1,000 \times 10^{-15}$ (10^{-15} = 1 femtometer),= 10^{-12} meters. As a result, the nucleus will decay to the ground state (ground state is the condition of an atom, ion, or molecule, when all of its electrons are in their lowest possible energy level i.e., not excited. Ground state also means that all electrons fill the lowest energy orbits.) by emitting one or more gamma-ray photons, Figure (5.8).

Figure (5.8): Comparison between the three isotopes in penetration

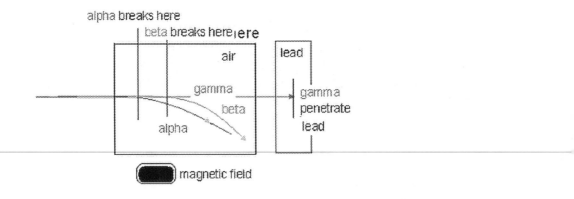

Alpha and beta decays can change the nucleus structure of an atom, i.e., the atom will be in an excited state after the decay is completed> After alpha and beta decay, gamma decay will take responsibility by releasing energy until the atom reaches the ground state, Figure (5.9).

Figure (5.9): staging of radioactive from Beta decay to Gamma decay

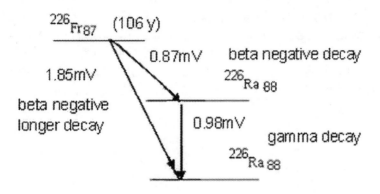

We shall show a full staging of energy and atomic numbers for the four decays; alpha, beta negative, beta positive, gamma in Figure (5.10).

Figure (5.10): Staging of alpha, beta negative, beta positive, and gamma decay.

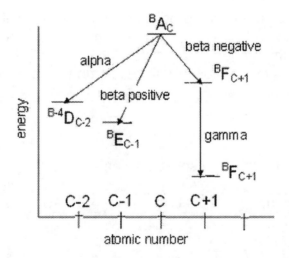

5.5.2 Half Time

Radioactive atoms (unstable isotopes) are steadily disappearing, and are being replaced by stable atoms. The decay from parent to daughter occurs at a certain rate. The rate of decay can be expressed as the time it takes for half of the weight of the parent to convert to its daughter. For example, the half time of beryllium 11 is 13.81 second, and let's start with 20 grams of ^{11}Be.

After 13.81 seconds, the weight if beryllium becomes 10 grams, and the rest is converted to boron - 11. Another 13.81 seconds, we will have 4 grams of beryllium -11, and 2 grams after 13.81 seconds more, and so on, until the beryllium -11 is vanished. We are talking about Be - 11 not Be - 9.012 which is stable. All radioactive elements disintegrate according to their specific half time, or half life. The half time of a radioisotope is the time required for half of the initial mass number of an atom to disintegrate. The following formula represents the decay time and the atomic mass of an atom:

$$Ln \frac{M}{Mo} = -kt$$

Where M is the remaining weight, Mo is the initial weight, k is the rate of decay, and t is the time of decay. Solve the above equation, considering M/Mo equals 0.5, then ln o.5 = - 0.693. Therefore, the half time equation can be written as:

$$t_{\frac{1}{2}} = \frac{0.693}{k}$$

Using the above equations, one can predict how much of an element is left over after a certain time, and how much of the element originally existed. Therefore, the ages of the archeological history of a substance (material, rocks, etc) can be determined. Table (8) shows some element with their half time decay.

Table (8): Half time of some elements

Uranium	$^{226}U_{92}$	4.5×10^9	y
Radium	$^{226}Rn_{88}$	1602	y
Radon	$^{222}Rn_{86}$	3.82	y
Astaline	$^{210}At_{85}$	8.1	h
Francium	$^{223}Fr_{87}$	22	min
Ununbium	$^{285}Uub_{112}$	34	sec
Ununhexium	$^{293}Uuh_{116}$	5.3×10^{-2}	sec

Example: Assume there is 875 grams of radon for every 125 grams of radium as:

1000 grams of radium (initially) → 500 → 250 → 125 grams of radium

Therefore, 1000 -125 = 875 grams of radon occurred in three half times.

The proportion of mass equals to 875 divided by 125 which is equal to 7.

Thus 7 times 1602 years of half times suggests that 11214 years have elapsed since the 1000 grams of radium converted to 875 grams of radon.

5.5.3 High Voltage Pulses and Patients with Bacterial and Viral Diseases

Cells of bacteria and viruses are known for their rapid growth (faster than our bodies' cells). They are much more versatile and diverse than our cells. High pulses of 3000 voltage can electroporate our cells to include bacteria and viruses in their plasmas and clone them for faster growth, resulting in cancer. Electroporation depends on voltage amplitude (300 to 3000 volts), frequency and wavelength. Unfortunately, in industrial countries heavy industry can produce pulses of all kinds, and theses pulses can travel hundreds of miles to other installation. Eletopolation of our cells' plasmas, incorporating bacteria, is shown in Figure (5.11). In the figure, bacteria enter the plasma, and increase and accelerate the growth of our cells.

Patients with bacterial and viral diseases should avoid exposure to industrial facilities, such steel plants, where arc furnaces and welding machines produce high voltage pulses. Traction (tramways), lightning, and automatic speed controllers of large electrical machines produce high voltage pulses. Electrodes can be represented by bones, and metals used in surgeries, dentures, and eyes.

Figure (5.11): the Electroporation process

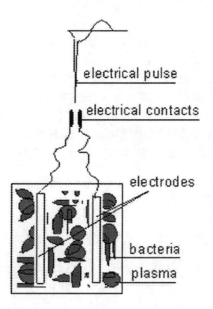

Chapter 6- Neuromuscular Electrical Stimulation (NMES)

Electrical impulses are being used for muscle contraction in the last few years. In the US, NMES devices are regulated by the Food and Drug Administration (FDA). Electrical stimulation for testing and training in exercise and sports has been discussed in many symposiums and conferences. The depolarization (change in a cell's membrane potential) of motor axons produces contractions by signals travelling from the stimulation location to the muscle (peripheral pathway). The associated depolarization of sensory axons sends a large stream into the CNS (the brain) and this can contribute to contractions by signals travelling through the spinal cord. Nerve stimulation has the advantage to be able to restore movement or reduce muscle atrophy and brain strokes.

6.1 How Neuromuscular electrical stimulation (NMES) Is Generated

Based on 2500 Hz being the best frequency to stimulate the muscle, the NMES causes the adaptation, i.e. training, of muscle fibers. Different types of pulses can activate different types of skeletal muscle fibers. The activation depends on the patterns of NMES shape and duration. These patterns cause different responses to different fiber types. Different patterns can be used for training, therapy, or cosmetic repair (such as muscle toning in the body, and micro-lifting of the face).

Figure (6.1) explains how to create the proper wave shape of a pulse, impulse or momentary steady voltage, and the way to make alternating polarity of the wave shape. The operation is as indicated below:

a) A single pulse, repetitive pulse, or broad band pulse is applied through resistor R1 to the transistor Tr1.
b) The collector of Tr1 now drops to near zero, and the voltage develops across R2 and passes the current to Tr2 which turns on.
c) When Tr2 opens, the current passes from the positive terminal of the supply through Tr2 to the ground (negative).
d) The pulsed wave at the output can be used for treatment.

Figure (6.1): The generation of pulses

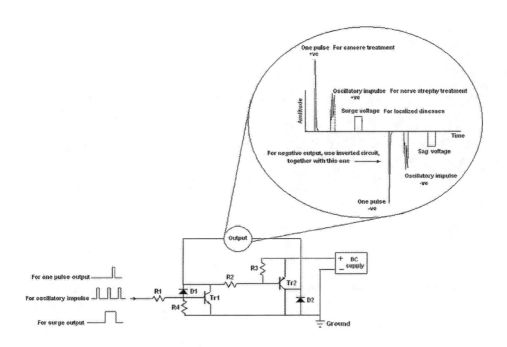

6.2 How Electrical Pulses Work

The high frequency (2500 Hz) is used only for rigid muscle treatment like those in the arms and legs. Muscles in the spinal cord can accept between 130 up to 750 Hz. In general, muscle fibers respond to frequencies from 20 Hz to 80-150 Hz.

At rest (you are sitting in your chair, arms resting on your lap), the voltage of the cells is between -55 to -60 mV (milliVolts). This is because the sodium is outside the cell, and potassium is inside (potassium has more negativity than sodium it has a higher number of electrons in one atom.

When the electrical pulse passes through the nerve, it prefers to travel through the outer surface of the nerve cells, where there is more sodium than potassium. Since the pulse (current) carries electrons, it will have the tendency to ionize the sodium, i.e. to convert the sodium Na to Na$^-$.

Figure (6.2): An electrical wave shape passing through a nerve due to changing ions in sodium and potassium

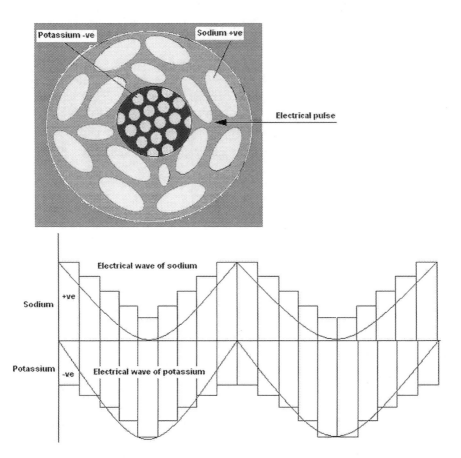

Sodium ions gradually enter the fiber through the leaky channels, and the voltage rises until a threshold level of about -40mV is reached. At this point, other channels are activated and there is a sudden rush of positive ions into the fiber. Sodium and calcium ions rush in while potassium ions flow out of the fiber. This causes an alternating current to flow in the nerve, Figure (6.2).

If a muscle fiber is exposed to 2500Hz, the vibration of the muscle would be 100,000 for 40 seconds, during which the reserve energy will disappear due to intense contraction. At this point, the muscle will stop responding to the stimulus, and another pulse is needed to carry on the treatment. When the muscle stops responding to the stimulus, the blood transports food and oxygen to compensate for the lost energy. Therefore the heart pulses and pressure will increase. If the heart pressure reaches the border line (140 for systolic and 90 for diastolic) the frequency must be dropped below 2500Hz, or the frequency of 2500 Hz is modulated (the amplitude and wave length are modulated). The number and shape of a muscle vibration depends on two features:

a) The type of electrical pulse – single pulse, oscillatory pulse, or broad band pulse, Figure (6.10)
b) Slow or fast twitch muscle fiber. This will be covered in the following section.

6.3 Stimulation of Slow and Fast Twitch Muscle Fibers

Muscle fiber types are broken down into two main types: slow twitch (Type I) muscle fibers and fast twitch (Type II) muscle fibers. Fast twitch fibers can be further classified into Type IIa and Type IIb fibers.

Slow Twitch (Type I): These are also known as red muscle fibers. They are responsible for long-duration, low-intensity activity such as walking or any other aerobic activity.

Fast Twitch (Type II): These are known as white muscle fibers (divided further into IIa and IIb). They are responsible for short-duration, high-intensity activity. Type IIa fibers are designed for short-to-moderate duration, moderate-to-high intensity work, as is seen in most weight-training activities. Type IIb fibers are built for explosive, very short-duration activities such as Olympic lifts.

Muscle contraction depends on the muscle type. For example, slow twitch muscle has more liquid and salinity, and therefore has better conductivity to electrical pulses. As a result, slow twitch muscle are recruited before the fast twitch muscle.

Frequencies of 40 Hz provoke muscle vibration. As the frequency increases, the vibration also increasesspeed, but more fatigable, muscle fibers are recruited. When the electrical stimulus reaches the nerve, it first penetrates the most peripheral and superficial, and then goes deeper and slower. The stimulation starts with slower twitch fibers and then the fast twitch fibers. Thus, both fibers are recruited successively, Figure (6.3).

Figure (6.3): The stimulation of Slow and Fast Twitch Muscle Fibers

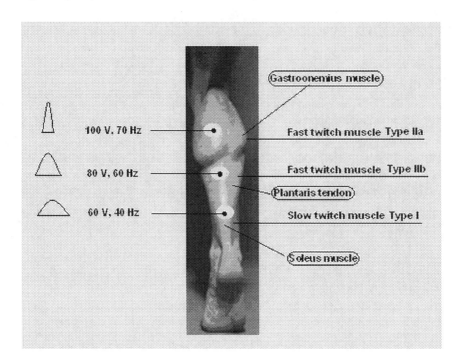

For better stimulation, electrical electrodes are generally placed on the motor points. A motor point is a macroscopic point, and there is one in each muscle belly. It is the point where the nerve crosses the muscle fascia to divide into millions of axons that terminate at their own end plate. This point is usually located in the middle of the muscle belly and coincides with the most prominent part of muscle. In a stroke patient, this point is identified by moving the electrode over the zone where we believe the motor point is located, until we achieve an optimal contraction.

The longest pulse time possible is selected (between 350 and 450 μs (microsecond), depending on the machine) so that the stimulation will not be painful. Different frequencies are also used in different muscle nodes (points). The pulse amplitude must not be painful; it has been observed that there is a limit to the amplitude. A technician must be trained for best results so that the patient does not feel pain.

In patients with strokes, impulse (pulses) are created and adjusted so that the "off time: between each train must be double (or more) the "on time", because the paretic (partial paralysis) muscles need greater time for recovery.

Chapter 7- Electrical Shocks by DC and AC Current

Both DC and AC currents are capable of killing. The current is more responsible for this than the voltage. It is the passage of the current that matters. For example, the current passing between two fingers of the same hand can burn, but the current passing between the two hands of a person could kill. Less than one amp is enough to kill.

In general, there are three types of currents: DC, AC and a mixture of DC and AC (even harmonics) as shown in Figure (7.1).

Figure (7.1): Types of electrical currents

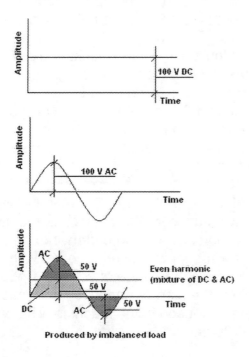

Even harmonic currents can occur with a short circuit of one phase to the ground, arc furnaces, imbalanced loads, capacitor banks, nonlinear loads, DC/AC converters, thyristorised controllers, and load disturbances.

Because DC moves with continuous motion through a conductor, it has the tendency to cause muscular tetanus (a state of continuous muscular contraction) quite readily. An alternating current (AC), it alternately, reverses direction of motion and provides brief moments of opportunity for a troubled muscle to relax between alternations. Thus, DC is more dangerous than AC. Even, a 60 volt DC battery is much more uncomfortable than a 120 AC coming out of the wall socket. However, an AC current nature has a greater tendency to throw the heart's pacemaker neurons into a condition of fibrillation, whereas DC tends to just make the heart frozen or stand still. Once the shock current is halted, a "frozen" heart has a better chance of regaining a normal beat pattern than a fibrillating heart. This is why "defibrillating" equipment used by emergency medics works: the jolt of current supplied by the defibrillator unit is DC, which halts fibrillation and gives the heart a chance to recover.

7.1 Defibrillation

Defibrillation is a common treatment for life-threatening cardiac arrhythmias, ventricular fibrillation, and pulseless ventricular tachycardia. Cardiac arrhythmias mean that the heart beats too quickly, too slowly or with an irregular pattern. If the arrhythmia is serious, the heart needs one of two devices implanted under your skin: a cardiac pacemaker or an implantable cardioverter defibrillator (ICD). An ICD is a device that monitors heart rhythms. If it senses irrigular rhythms, it delivers shocks (pulses). Many ICDs record the heart's electrical pulses when there is an abnormal heartbeat. This can help the doctor plan future treatment. Automatic External Defibrillators (AED) should be used within the first 3-5 minutes after a collapse. Currently less than 5% of victims survive largely because a defibrillator does not arrive in time. An AED can increase the survival rate to 50% and in some situations, like on commercial airlines, around 70%.

The electrical circuit diagram of the defibrillator is shown in Figure (7.2). It contains a high voltage DC power supply (1000-6000V). In the figure, a battery is drawn as a symbol as no conventional battery can produce such a high voltage. Usually, an AC power supply is used together with a rectifier and a step up transformer to produce that range of voltage. Both switches S1 and S2 operate simultaneously at the start. A capacitor of 16 microFarad and an inductor of low resistivity are also used in the machine. The patient's heart represents the resistance (50-100 ohms).

The operation starts with turning S1 and S2 on (they close and open simultaneously). The capacitor charges up to 4000 V and stores the energy until the switch S3 turns on. The capacitor discharges the pulse to the heart through the inductor which limits the current from a sudden shoot, and the patient resistance damps the current. An additional variable resistor can be connected in series with the inductor to change the voltage amplitude.

The Figure shows the wave shape of the pulse, called the "Lown" waveform, which is the typical discharge pulse of the defibrillator. The wave form width is about 5 ms in the positive region and 2.5 ms in the negative. The peak current could reach 20 A. The Lown wave shape can be reversed by reversing the poles of the battery or by reversing the poles of the rectifier. In some cases, reversed voltage is used in heart transplant and in the recovery period.

Figure (7.2): A defibrillator with its components and output wave shapes

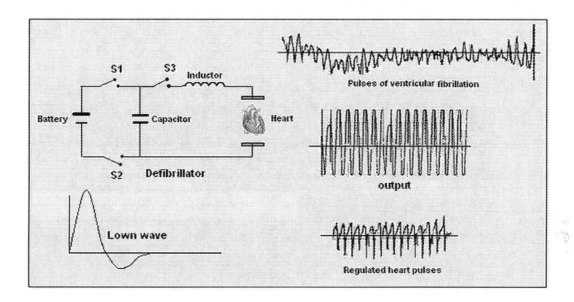

7.2 Electrocution

Electrocution is a type of electric shock that can stop the heart from beating, resulting in death. A sudden accident can be caused by an electric shock, and a deliberate execution by means of an electric shock is sometimes used, such as an electric chair.

Fibrillation can occur from shocks of small currents of a 70-700 mA capacity. Fibrillation is reversible by using defibrillator. However, currents above 1 A can cause devastation to the heart, next is the brain.

Electrocution depends on the following parameters:

a. Voltage: The voltage above 600 V can break the dielectric of the human body, and cause burns or stops the heart. Humidity decreases the resistance of the body, which is 500 Ohms at 50% humidity.
b. Current: High current above 1 A can kill. It depends on the resistance of the body.
c. Duration: The duration of the frequency of the shock has a large impact on the life. The larger the duration, the more likely it is fatal.
d. Frequency: High frequency can penetrate deeper, and can overcome resistance.

e. Pathway: A pathway from one finger of the same hand is different that from one hand to another.

f. Type of current: Shocks can be of DC, AC, or a mixture of both, all of which were discussed earlier in this book.

7.3 Electric Shocks in Medicine

Electric shock can also be used in medicine for treatment as in the following cases:

- Electroconvulsive Therapy (ECT) is a procedure in which electric currents of high pulses are passed through the brain, Figure (7.3). ECT triggers the brain, and possibly changes the brain chemistry or neurotransmitters that reverse actions of mental illness such as seizures and severe depression and bipolar depression. Much of the problems associated with electroconvulsive therapy are based on early treatments in which high doses of electricity were administered without anesthesia, leading to amnesia (memory loss), fractured bones and muscle strains.

Figure (7.3): Treatment of a patient with ECT

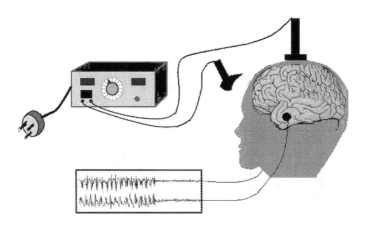

- Electrical shocks are used as a surgical tool for cutting or coagulation. An "Electrosurgical Unit" (or ESU) uses high currents (10 amperes or above) at a high frequency (500 kHz or above). The current amplitude can be varied for better results.

With the cutting modes, the electrode is held over the tissue and a small continuous arc is produced between the electrode and the tissue. This in turn causes rapid cell destruction followed by the complete explosion of the cell. It 'blows away' cell debris achieving a clean cut. Should the electrode be placed directly on the tissue and the cut waveform used, some coagulation will be achieved. This effect is called desiccation.

Within coagulation mode, a discontinuous (interrupted) current is used, driven by a high voltage.

- As a treatment for fibrillation or irregular heart rhythms (explained earlier in this book).

- As electroanalgesia for many types of pains. Transcutaneous Electrical Nerve Stimulators (TENS) are being used in hundreds of clinics for treating various types of conditions, such as low back pain, myofascial and arthritic pain, sympathetically mediated pain, bladder incontinence, neurogenic pain, visceral pain, and postsurgical pain. Figure (7.4) shows an image of TENS

Figure (7.4): TENS image

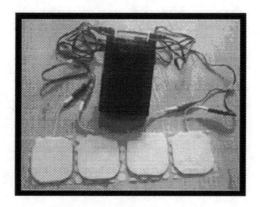

- As an aversive punishment for the conditioning of mentally handicapped patients with severe behavioral problems. This method is highly controversial and is employed at only one institution in the United States, the Judge Rotenberge Educational Centre. The institute also uses electric shock punishments on non-handicapped children with behavioral problems. Whether this constitutes legitimate medical treatment or abusive discipline is currently the subject of litigation. The Center administers 2-second electric skin shocks to residents using a Graduated Electronic Decelerator (GED), which was invented to administer the skin-shocks by remote control through electrodes worn against the skin. Most often, the shocks are initiated manually by the staff. Automatic punishment is also used by forcing the patient to sit down on a cushion; if they stand up, they are automatically shocked.

7.4 The General Principles of Somatosensory Evoked Potentials

Evoked potentials are the electrical signals generated by the nervous system in response to sensory stimuli. Auditory, visual, and somatosensory stimuli are used commonly for clinical evoked potential studies. Somatosensory evoked potentials (SEPs) consist of a series of waves that reflect sequential activation of neural structures along the somatosensory pathways. While SEPs can be elicited by mechanical stimulation, clinical studies use electrical stimulation of peripheral nerves, which gives larger and more robust responses.

The stimulation sites typically used for clinical diagnostic SEP studies are the median nerve at the wrist, the common peroneal nerve at the knee, and/or the posterior tibial nerve at the ankle. The recording of SEPs to the stimulation of the ulnar nerve at the wrist is useful for intraoperative monitoring when the mid cervical spinal cord or parts of the brachial plexus are at risk. Recording electrodes are placed over the scalp, spine, and peripheral nerves proximal to the stimulation site.

Evoked potentials are generated in different shapes. The most common shapes are:

a- Successive enlarged wave lengths as shown in Figure (7.5-a), or
b- Successive enlarged amplitudes, Figure (7.5-b)

Figure (7.5): Wave shapes of evoked potentials

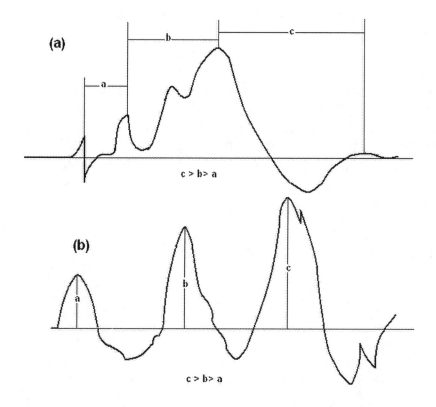

7.5 Laser Evoked Potentials (LEPs)

Stimulation of the dorsal skin with brief laser impulses easily evokes brain potentials (laser evoked potentials, LEPs). This non-invasive technique is applied at the present time in diagnosing small fibre neuropathy. In order to record distinctive and repetitive evoked potentials from the surface of the skull, the skin of the dorsal surface of the hands and feet is stimulated with a thermal laser.

The generation of a pain elevated-related brain requires accurate onsets of stimulus of laser evoked potentials (LEPs) of strong, brief phases. Currently, LEDs are the most reliable and commonly used method for investigating pain in the brain.

For pain stimulation, a helium laser stimulator with a laser-beam wavelength of 1.8 mm and a beam diameter of 30 mm2 is used. Other wavelengths and beam diameters can also be used for skin layers between the dorsal and ventral layers. Stimuli consist of brief heat pulses with durations of 1.5 ms. Heat stimuli are applied to the dorsum of hands, legs, back and other movable parts of the body. The stimulus location was visualized by a red helium laser beam pointing onto the dorsum of the body. CO2 laser beams can also be used. To avoid possible tissue damage caused by heat accumulation in the layers of the skin, the following safety measures should be applied:

- Workers must wear protective goggles throughout the whole operation when laser stimuli are applied.
- To avoid tissue burns and damages, the location of stimulus application should be slightly shifted after each single stimulus, by stepwise displacing the optic lens of the laser using a computer-controlled stepmotor. This procedure assures that no spot on the skin is stimulated twice within succeeding trials of stimulation.
- To minimize risk of skin burns or skin irritation, the energy of laser stimulation should be kept below a maximal value of 600 mJ.

Chapter 8- Human and the Disturbance of Electrical Interference

A spark is triggered when the electric field strength exceeds approximately 4–30 kV/cm — the dielectric field strength of air at relative humidity is 10% or less. This may cause a very rapid increase in the number of free electrons and ions in the air. For example, oxygen (O) becomes O negative or positive. This temporarily causes the air to abruptly become an electrical conductor in a process called dielectric breakdown. Dielectric breakdown occurs when a charge buildup exceeds the electrical limit or dielectric strength of a material. The negatively charged electrons are pulled in one direction and the positively charged ions in the other. When electrons are removed from a nucleus, it becomes positively charged. When air molecules become ionized in a very high electric field, the air change from an insulator to a conductor. Sparks occur because of the recombination of electrons and ions. Lightning occurs when there is a buildup of charge on the clouds and on the ground it produces an electric field that exceeds the dielectric strength of air. Ionized air is a good conductor and provides a path where charges can flow from clouds to ground.

The dielectric strength of air is approximately 3 kV/m. Its exact value varies with the shape and size of the electrodes and increases with the pressure of the air.

There are several protective measures you can do to help minimize getting shocked:

- Shoes (soles) should not have metallic dividers or conductive synthetic materials.
- Relative humidity (RH) should not be less than 40%, because higher humidity makes things have better conductivity.
- Carpet and fabric should be sprayed with antistatic spray, if the relative humidity is getting high.
- Place grounded surfaces over exposed metal or conductive surfaces.

8.1 Electric Human

There are reports of rare people, "Electric Humans," who build up high voltage on their bodies and suffer the continuous problem of electrical sparks. Their sparking occurs regardless of footwear, clothing, humidity, or even motion! Electric people are forever getting zapped when they touch others, or when they touch large metal objects like door knobs. For the rest of us, "static electric" sparking can only occur after we walk across certain carpets, or when we wiggle around while sitting on certain chairs. For the rest of us, the problem vanishes when the humidity is high, when we go barefoot when we avoid wool or nylon sweaters/pants, when we avoid plastic seats, etc. But the bodies of "Electric Humans" instead seem to become electrified all the time, all by themselves. It is a common phenomenon which is constantly reported by members of the public, but which is denied by science because it "goes against theory. Sparks can be irritating and their cause sometimes seems mysterious". "Electric humans" claim that they can turn on the TV by holding the plug, move toys without batteries, and breathe out charged air. It is alleged that there is an anomalous phenomenon where the person seems to turn off (or sometimes on) street lights or outside building security lights, when passing near them.

One explanation to the phenomenon of the electric human is that "if a person were to constantly be breathing out negative ions (charged air molecules), then unless their body was electrically grounded to the earth, they would rapidly accumulate a strong positive charge-imbalance on their body; an imbalance which is equal and opposite to the charged air being breathed out". But why would our lungs be producing electrified air? One possibility: when microscopic bubbles burst, the natural surface-charge of the water will cause the spray of tiny droplets to have a negative charge. If the liquid on the inner surface of alveoli in our lungs should be full of micro-bubbles, then our lungs might become VandeGraaff generators (http://amasci.com/emotor/zapped.html#ehum).

8.2 Sparks and Health hazards

Exposure to an arc-producing device can pose health hazards. In a closed space such as a classroom or home, the continuous arc formation will ionize oxygen, nitrogen and other gases, which then re-form into reactive molecules such as ozone (O_3) and nitric oxide (NO). Oxygen and nitrogen will also be ionized (O^-, O^+, N^-, N^+) These free radicals can be damaging to the mucous membranes of people near the spark gap. Plants are also susceptible to ozone poisoning.

Sparks can play a big role on the acidity and alkalinity of the body. Sparks ionize the fluids of the body which can be too acidic (high level of H^+) or too basic (high level of OH^-). The body buffers the level of both acidity and alkalinity by producing carbonic acid (H2CO3) or bicarbonate (HCO3). If there is extra hydrogen in the body, the PH will drop. The pH level has profound effects on all body chemistry, health and disease. All regulatory mechanisms (including breathing, circulation, digestion, hormonal production) serve the purpose of balancing the pH level by removing caustic metabolized acid residues from body tissues without damaging living cells. If the blood pH level drops below 6.8 or increases above 7.8, cells stop functioning and the patient dies.

Arcs can also produce a broad spectrum of wavelengths spanning the visible light and the invisible ultraviolet and infrared spectrum. Very intense arcs generated by means such as arc welding, can produce significant amounts of ultraviolet rays which is damaging to the retina of the observer. These arcs should only be observed through special dark filters which reduce the arc intensity and shield the observer's eyes from the ultraviolet rays.

Sparks can be generated by many electrical devices and systems, including:

- Electric tramways
- Thermostats, fridges, freezers, heating and cooling systems
- Electric motors, speed drives, accelerators
- Robotic systems
- Capacitor banks, and power factor correctors
- UPS, rectifiers, inverters, and voltage stabilizers
- Arc furnaces, and the steel industry
- Transmission lines, the corona effect, and HVAC to HVDC converters using thryristors and diodes
- Ignition systems on cars and motorbikes
- Dimmer switches, ballasts and starters in discharge lamps
- Substation, switch gears, and circuit breakers.

8.3 Biological hazards

The biological effect of electromagnetics can cause heating in the human body similar to the burns that would happen inside a microwave oven. For example, standing or touching an antenna while a high-power transmitter is in operation can cause severe burns.

This heating effect is measured in watts per kilogram of weight, and varies with the power and the frequency of the electromagnetic power. The IEEE and many national governments have established safety limits for exposure to various frequencies of electromagnetic energy. IEEE journals publish research and applications in advances in antennas that is theoretical or experimental. Design and development is included in this focus of antenna advancement. The radiation of an antenna depends on the wave length of frequency, the width of the antenna, the type of the antenna, and closeness to the antenna. Figure (8.1) shows two types of antennas and the factors that affect the radiation.

Figure (8.1): The radiation of antennas depends on several parameters

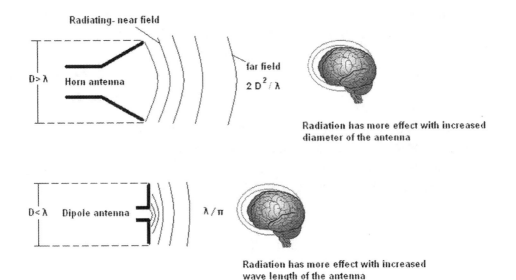

8.4 EMI and DNA fragmentation

1) DNAase (an enzyme that destroys DNA) leaking through the membranes of lysosomes (small bodies in living cells packed with digestive enzymes) explains the fragmentation of DNA seen in cells exposed to mobile phones, computers, power lines and domestic wiring. When this occurs in the germ cells (the cells that give rise to eggs and sperm), it reduces fertility and predicts genetic damage in future generations.

   As we mentioned before that DNA fragmentation could lead to cancer. Cancer cells divide rapidly, the daughter cells divide before they have reached maturity. Electromagnetic waves, pH, temperature, and some drugs (enzymes) may affect the rate of division. When DNA is fragmented, cells stop dividing; they stop at a point late in the stage G1, read stages of cell division in other literatures. The cell division will not reach the stage S which is the stage when the DNA is replicated for the next division, and the chromosomes become double stranded. Thus, the DNA stops replicating, and the cell will not enter the G2 stage and proceeds in to cell division. Cells will not divide again and stop in the G1.

2) Defects of the DNA in the mitochondrial genome advance the cell to be aged. The healthy mitochondrial genome is a closed circular DNA

molecule and it encodes 22 tRNAs, two rRNAs and 13 polypeptides of the electron transport system. From two to ten copies of the genome can be found in each mitochondrion. The mitochondrial genome is susceptible to mutation damage because the circular DNA is fragmented (damaged), therefore, it lacks histone protection and DNA repair systems. These mutations include duplications of regions of the mitochondrial genome, point mutations and deletions of large segments of the genome. The mtDNA (mitochondrial DNA) deletion mutations accumulate with age and are commonly detected in post-mitotic tissues, such as the brain, heart and skeletal muscle, which rely heavily on oxidative metabolism.

8.4 EMI and Signals from the Brain to Other Organs

The human brain is a complex organ that controls our body, receives and transmits electrical signals, and analyzes and stores information. It is the organ that allows human to see, hear, smell, think, taste, move, and feel. The brain generates electrical signals that interact with chemical particles in the body. The result is transmitted through wires (nerves) to motor other organs as required. Figure (8.2) represents a simulation between the brain (generator) and the body's organs, and an interruption to the signals due to the EMI

Figure (8.2): EMI interrupts feedback signal between brain and organs.

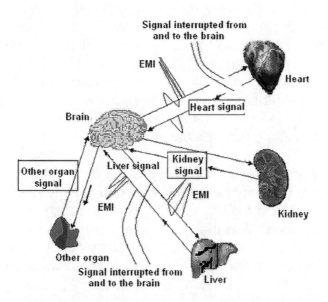

8.5 Electrical Charges and Metabolism

In metabolism, the compound accepts or donates electrons in redox (reduction and oxidation) reactions. Since NAD^+ is a cation, it is involved in redox reactions, taking electrons from one reaction to another. The coenzyme is therefore found in two forms in cells: NAD^+ is an oxidizing agent and NADH is a reducing agent, Figure (8.3).

Figure (8.3): The reduction and oxidation between NAD^+ and NADH

NAD+ and NADH have emerged as novel, fundamental regulators of electrolytes particularly calcium homeostasis. It appears that most of the components in the metabolic pathways of NAD+ and NADH, including poly (ADP-ribose), ADP-ribose, cyclic ADP-ribose, O-acetyl-ADP-ribose, nicotinamide and kynurenine, can produce significant biological effects. Numerous studies have suggested that NAD+ and NADH mediate multiple major biological processes, including calcium homeostasis, energy metabolism, mitochondrial functions, cell death and aging. Studies also proposed that NAD+ and NADH are fundamental mediators of brain functions, brain senescence and multiple brain diseases.

NAD+ and NADH play an important role in removing acetyl groups that are commonly transferred to conenzyme A (CoA) or from acetyle-CoA. Acetyl-CoA is an intermediate both in the biological synthesis and in the breakdown of many organic molecules. CoA is the final stage of the process of glycolysis in the cytoplasm where two stages of glycolysis take place. Energy investing with NAD+ is one product, and energy harvesting with ATP is another product, Figure (8.4).

Figure (8.4): Glycolysis of glucose producing CoA

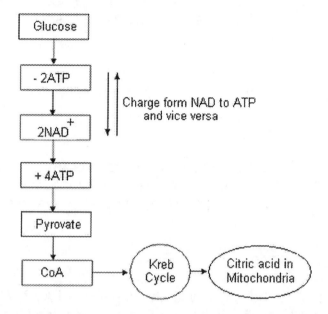

Glycolysis in the cytoplasm

At the beginning of the Krebs cycle, protein and fat need oxygen to burn them. This oxidation process is accomplished by the NAD which takes the hydrogen from foods to ease the oxidation. The NAD becomes oxidized, and the NADH is reduced (electron-charge rich).

8.6 Electromagnetic Fields and Aging

Oxidative stress due to radical components can damage mitochondrial DNA (mtDNA) and cell DNA. DC pulsed electromagnetic fields at under 10 Hz could enhance cell respiration mechanisms and lower free radical waste levels to sufficient extent for the organism to perform as well as it did at 1/2 its age. Only sleeping in a 10 Hz field would provide genetic effect as described by NASA. A field above 10 Hz could damage all types of DNA and cell mitochondria.

Electromagnetic radiation can be classified into ionizing radiation and non-ionizing radiation, based on whether it is capable of ionizing atoms and breaking chemical bonds. Both electromagnetic types could be associated with three major potential hazards: electrical, chemical or biological.

The IEEE (USA), IEE (UK), Cigre (France), VDE (Germany), JIS (Japan) and many other international associations have established safety limits for exposure to various frequencies and voltages of electromagnetic fields. Here are some limitations:

- o For residential exposure – 1000 milli gauss (unit of measurement for magnetic flux density)
- o For occupational exposure – 5000 milli gauss
- o Power lines should not exceed 1000 milli gauss for 24 hours,
- o Magnetic fields in schools, daycares, and play grounds should not exceed 2-3 milli gauss (Sweden).

In Europe, loads of non-linear types (inductive or capacitive, but not resistive) must have filters built into them at the factory. As a result, American computer manufacturers must put an inexpensive RF (radio frequency) filters in the computers they export to Europe.

High frequency can cause severe cell injuries of varying degrees, and senescent apoptotic or necrotic cell death. Apoptosis can be induced either by a stimulus, such as irradiation or toxic drugs, or by removal of a repressor agent.

The reversal of aging could be achieved by:

- Reduced oxidative phosphorylation with the depletion of ATP. With high frequencies, the phosphorylation is not completed and thus the pyrovate converts into lactic acid, causing the pH level to reduce, the synthesis of ATP to slow down, and the P/O ratio (phosphate per oxygen ratio) to reduce. Therefore, increased oxidative phosphorylation and reduced frequencies below 10 Hz could reverse aging.
- Cellular edema which is caused by changes in ion and water flows. K^+ channel-blocking Ba^{2+} ions, and Cl^- induce osmotic swelling in cells. Ions are caused by high frequencies.
- The generation of reactive oxygen is responsible for the toxicity of Mitochondrial and cytoskeleton alterations. The cytoskeleton is a network of protein filaments that are responsible for the movement and support of the cell. Figure (8.5) shows mitochondrion and cytoskeleton in a human cell.
- DNA damage can be repaired. In human cells, both normal metabolic activities and environmental factors, including UV light and radiation can cause DNA damage, resulting in as many as 1 million individual molecular lesions per cell per day. (Lodish H, Berk A, Matsudaira P, Kaiser CA, Krieger M, Scott MP, Zipursky SL, Darnell J. [2004], Molecular Biology of the Cell, p963. WH Freeman: New York, NY. 5th ed). Many of these lesions cause structural damage to the DNA molecule and can alter or eliminate the cell's ability to transcribe the gene that the affected DNA encodes. DNA damages can be recognized by enzymes, and they can be correctly repaired if redundant information, such as the undamaged sequence in the complementary DNA strand or in a homologous chromosome, is available for copying.

Figure (8.5): The mitochondria and cytoskeleton in a human cell

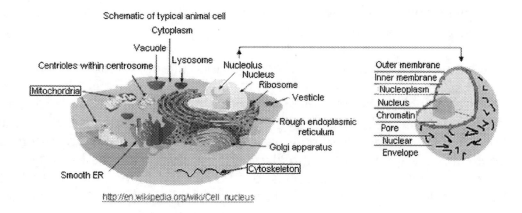

Chapter 9- Magnetic Field Therapy

Magnetic field therapy diagnoses and treats both physical and emotional pain; it relieves symptoms and delays the disease. Magnets and electromagnetic therapy equipment are now being used to remove pain, help the healing of broken bones, and counter the effects of depression.

Recently, engineers found that magnetic fields affect the human body physically and emotionally in both positive and negative ways. They discovered that the body's cells are affected by magnetic fields, which are produced by nature's magnetic field (the earth's magnetic field and the sun's burst magnetic fields), and man made magnetic fields (created by electrical equipment). Magnetic fields can also be generated inside the body by chemical reaction within the cells and ionic currents of the nervous system. For example, oxidation and reduction in the body is accompanied by ionic reaction. Similarly, acids and bases can generate currents between ionic charges. The following facts have been found helpful in magnetic field therapy:

a. Negative magnetic fields have a beneficial effect on living organisms, whereas positive magnetic fields have a stressful effect. A prolonged exposure to a positive magnetic field could increase acidity (low PH), and consequently reduces cellular oxygen supply. The reduction of oxygen does not help patients with cancer (Warburg Effect). Positive magnetic fields can increase pain because their interference with metabolic function. The biological value of oxygen is increased by the influence of a negative electromagnetic field. The field causes the negatively charge (DNA) to "pull" the oxygen out of the bloodstream and into the cell. Negative fields increase the PH to an alkalinity level.

b. It was found that weak electric currents promote the healing of broken bones. However, electromagnetic interference from power lines and home appliances (they are not weak electric currents) can pose a serious hazard to human health.

c. Kyoichi Nakagawa, M.D., Director of the Isuzu hospital in Tokyo, Japan, believes that the time people spend in buildings and cars reduces their exposure to natural geomagnetic fields of the earth, and may interfere with their health. He calls the condition that they subsequently must suffer from "magnetic field deficiency syndrome", which, he says, can cause headaches, dizziness, muscle stiffness, chest pain, insomnia, constipation, and general fatigue.

d. Proper electrical pulses can kill cancer cells. Electrofulgeration destroys a cancer cell by burning it with an electrical current. Radio frequency burning can reduce the tumor by inserting needles connected to a radio frequency electric current. The current is turned on for 2 to 3 minutes to burn and kill the tumor cells.

e. Electrolytes (Na^+, K^+, Ca^{2+}, Mg^{2+}, Cl^-, PO_4^{3-}, and HCO_3^-) can be elevated (hyper) or depleted (hypo) du to the effect of electrical disturbances such as pulses and impulses. So, electrical pulses can be used to control and adjust the electrolyte level, and cosequently remove the stress on the kidneys.

f. Electromagnetic waves and pulses can have an effect on the shape of hyper and hypo electrolytes. For example, hyperkalemia (extra potassium) has a "tent shape" in normal conditions. When the body is exposed to high electrical pulses, the hyperkalemia changes its pattern to a T or U shape, which are the characteristics of hypokalemia. When untreated, hypokalemia may lead to arrhythmia which is related to heart disease.

g. The electroporation process in surgeries can be used for the treatment of dentures, eyes, and bones. Electrodes can be represented by bones, and metals in such surgeries.

9.1 Pulsed Electro Magnetic Field Therapy – PEMF

Pulsed Electro Magnetic Field Therapy (PEMF) is used for the treatment of many ailments. The pulses range between 3 and 1000 Hz, described as follow:

a. 200 Hz can improve blood circulation
b. 25 Hz can be used to detoxify blood and radical components in the body
c. 3 Hz can be used to relax the body and mind; improves sleep and reduces nervous tension
d. Automatic eight minute cycle with the reversal of magnetic polarity every two minutes can be done
e. Active pulsing for nine seconds, then three seconds off throughout the eight minute cycle prevents adaptation or accommodation as may occur with systems using a constant, non-varying magnetic field. Figure (9.1) illustrates the above frequencies

Figure (9.1): Illustration of the above frequencies

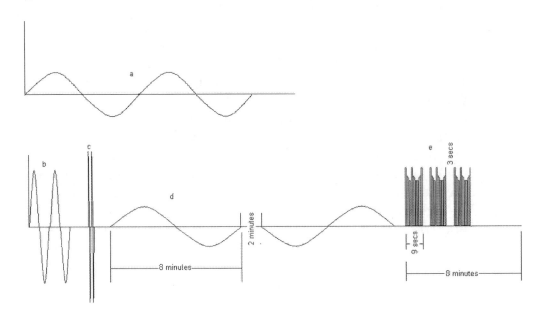

9.1.1 PEMF for Pain Relief

There are many types of pains. For example, burning, pinching, freezing, torturing, and terrifying are types of pain. Pain can be steady and constant, and relates to a sensation that aches. If you feel pain it hurts, you feel discomfort, distress and perhaps agony, depending on the severity of it. Pain is said to be caused by the disruption of blood or the lack of oxygen. Blood pressure is controlled by the heart. Oxygen flow is affected by energy and electromagnetic waves. The heart can not work without electrical pulses (see the section of the heart in this book), and the oxygen is controlled by many factors, all of which are related to electrical energy.

For example, PEMF of high frequencies could take amino acids to bits (radical nitrogen, N⁻), and the sodium in the body will combine with the radical nitrogen to form sodium nitrate and sodium nitrite. Sodium nitrate has the following formula:

$$\underset{\overset{|}{\underset{-O}{\overset{O}{\parallel}}}}{N^+}\diagdown O^-\ Na^+$$

Both sodium nitrate and sodium nitrite form carcinogic molecule nitrosamines – human carcinogens known to cause DNA damage and increased cellular degeneration. Studies have shown a link between increased levels of nitrates and increased deaths from certain diseases, including Alzheimer's, diabetes mellitus and Parkinson's, possibly through the damaging effect of nitrosamines on DNA. Low frequency is used for controlling heart pulses and radical elements.

9.1.2 PEMF for Osteoporosis Treatment

Osteoporosis is a bone disease that leads to an increased risk of fractured bones. In osteoporosis the bone mineral density (BMD) is reduced because of low dietary calcium and/or phosphorus, magnesium, zinc, boron, iron, fluoride, copper, vitamins A, K, E and C (and D where skin exposure to sunlight provides an inadequate supply). Radical calcium contributes to osteoporosis. PEMF can be used to control the amount of minerals.

Calcium is an important element to keep bones and skin healthy, and prevent early aging. All acetic acids in the body interact with calcium. Acetic acids are components of fatty acids, and they can be used for energy by most types of cells. Electrical and magnetic waves can affect both acetic acids and calcium, and produce something different as per the following equation:

$$2CH_3COOH + Ca \longrightarrow Ca(OOCCH_3)_2 + H_2$$

The product is calcium acetate and hydrogen. Calcium acetate has two radicals: oxygen negative and calcium positive. Free radicals can cause aging, heart disease, hypertension, stroke and cancer. It has been claimed that free radicals can cause headaches, dry skin and high blood pressure, which are all attributes of aging. PEMF can be used to reverse the radicals.

9.1.3 PEMF and Fibromyalgia

Fibromyalgia is a disorder characterized by widespread musculoskeletal pain accompanied by fatigue, sleep, memory and mood issues. Researchers believe that fibromyalgia amplifies painful sensations by affecting the way your brain processes pain signals.

Fibromyalgia is associated with low oxygen in tissues. This is a result of issues in the mitochondrial electron transport chain (oxygen metabolism).

Pulsed electromagnetic field (PEMF) therapy has proven effective on pain for 5 decades, particularly in Eastern European research.

The metabolic process in the body converts energy from nutrients into adenosine energy (ATP). In the process, the sugar is converted to pyruvate which involves the redox (reduction and oxidation) reaction. The pyruvate enters the mitochondrion to break into glycolysis. Oxidation is accomlished in the mitochondrion through the Krebs cycle. The product of this process is energy in the form of ATP (adenosine triphosphate), which is done by phosphorylation, NADH and $FADH_2$.

Pulsed electromagnetic field (PEMF) therapy can speed up the process of oxidation and reduction in the Krebs cycle. At the end of the glycolitic (pyruvic acid) PEMF boosts, electrons spark and help power the Krebs cycle to generate more of the ATP.

9.1.4 PEMF and Fracture Healing

The skeletons and teeth of vertebrate animals are composed of calcium phosphate, mainly (tricalcium phosphate) which is shown below:

Tricalcium phosphate

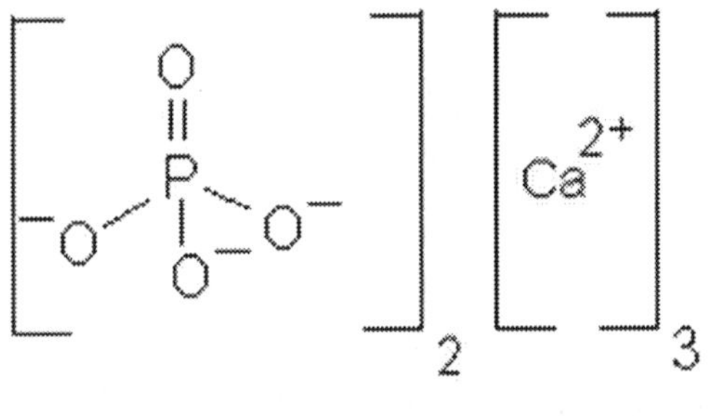

One can see that PEMF with a modified frequency can increase the tricalcium phosphate by controlling the ions of both phosphate and calcium. Triphosphate can also be used for DNA repair.

9.1.5 PEMF and Rheumatic Pain

Rheumatic pain or rheumatoid arthritis includes diseases like osteoarthritis, lupus, fibromyalgia, and tendonitis. A recent study showed that rheumatoid arthritis is caused by autoimmune disorders. A disturbance in the immune system could develop autoimmune diseases. The autoimmune system sees substances and tissues normally present in the body as enemies (pathogens), and attacks them. Autoimmunity is the failure of an organism in recognizing its own constituent parts as *self*, which allows an immune response against its own cells and tissues. Sometimes the immune system will cease to recognize one or more of the body's normal constituents as "self" and will produce autoimmune – antibodies that attack its own cells, tissues, and/or organs. This causes inflammation and damage and leads to autoimmune disorders. Three main sets of genes are suspected in many autoimmune diseases. These genes are related to:

- T –cells
- The major histocompatibility
- Immunoglobulins

In nearly all vertebrates, B and T cells are produced by stem cells in the bone marrow. T cells travel to and develop in the thymus gland. Both B and T cells recognize specific antigen targets. B cells complete their maturation in the bone marrow. Inside the thymus gland, T cells educate themselves to distinguish self cells from non-self cells. Figure (9.2) shows the chain of the immune system with T cells.

Figure (9.2): T cells are produced by stem cells (form my book: "Biochemistry of Aging")

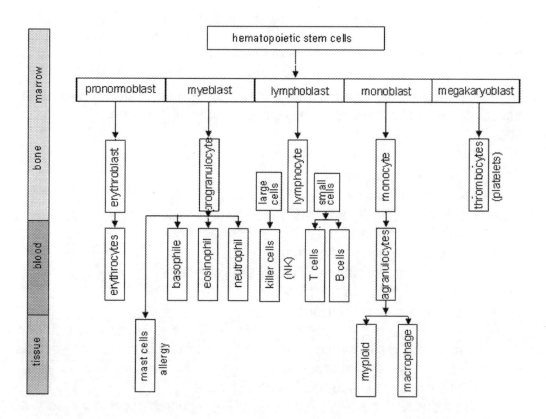

The interaction of nanosecond duration pulsed electric fields (nsPEFs) with biological cells, and the models describing this behavior, depend critically on the electrical properties of the cells being pulsed (http://eprints.soton.ac.uk/264453/).

Malignant human T cells were exposed to pulsed electrical fields of variable amplitudes and wave lengths. The conductivity of the cytoplasm and nucleoplasm was decreased dramatically after pulsing. This suggests that electropermeabilization occurred, resulting in ion transport from the cell's interior to the exterior. A delayed decrease in cell membrane conductivity after the nsPEFs possibly suggests long-term ion channel damage or use dependence due to repeated membrane charging and discharging.

In another study, an electrical model for biological cells of different patterns can be used to interact with intercellular structures of T cells, and restore the function of the immune system to its normality. Pulses of 10 nanoseconds and 500 kV/cm could achieve that, although some other cells would be subjected to apoptosis (cell death). Such an application has also been shown to reduce the growth of cancer cells.

9.1.6 PEMF and Stress Reduction

Cortisol is a hormone secreted by the adrenal glands, and it has been termed " the stress hormone". It is secreted into the bloodstream in higher level during the body's 'fight or flight' response to stress.

Higher and more prolonged levels of cortisol in the bloodstream (like those associated with chronic stress) have been shown to have negative effects, such as:

- Higher blood pressure
- Decreased bone density
- Lowered immunity and inflammatory responses in the body, slowed wound healing, and other health consequences
- Impaired cognitive performance
- Blood sugar imbalances like hyperglycemia
- Suppressed thyroid function
- Decrease in muscle tissue
- Increased levels of "bad" cholesterol (LDL) and lower levels of "good" cholesterol (HDL), which can lead to other health problems such as heart attacks and strokes
- The kidneys to producing hypotonic urine
- Shut down of the reproductive system, resulting in an increased chance of miscarriage and (in some cases) temporary infertility. Fertility returns after cortisol levels return to normal

It has been found that music therapy of low rhythm (low frequency) can reduce cortisol levels in certain situations. Experiments using pulsed electromagnetic fields of low frequency (PEMF) showed reduction in the physical response to the effects of daily stress in the whole body (http://www.magnopro-usa.com/research/Stress-and-PEMFs.pdf).

Research, on humans and animals, has shown that the PEMF alters stress responses by electromagnetic action directly on the nervous system, glands, cells, tissues and organs.

Chapter 10- The Electrical System of the Heart

10.1 The Heart

The heart of the human body is located slightly to the left of the middle of the thorax, behind the breastbone (sternum). It is enclosed by a sac known as the pericardium and is surrounded by the lungs. It weighs about 300-350 g in an adult. It consists of four chambers, the two upper atria and the two lower ventricles.

Animals have different heart beats than humans. For example, a whale's heart beat 8-10 times per minute, a seal 10 when diving and 140 times on lands, elephant 25, sparrow 500, hummingbird 1200 when hovering, and human 70. The pumping of the heart is called the Cardiac Cycle, which occurs about 70 - 72 times per minute. This means that each cycle lasts about eight-tenths of a second. During this cycle the entire heart actually rests for about four-tenths of a second.

The heart works as a pump moving blood around in our bodies to give food to every cell. The blood returns from end distribution points to the heart, and then sends it back through the lungs for reoxygenation. The output of the blood from heart is about 10 liters per minute and about 20 liters per minute for a trained athlete. The average output over a life time by the age of 70 years would be 2000 US gallons (7,570 liters) of blood per day.

10.2 The Structure of the Heart

The heart has 4 chambers. The upper chambers are called the left and right atria, and the lower chambers are called the left and right ventricles, Figure (10.1). A wall of muscle called the septum separates the left and right atria and the left and right ventricles. The left ventricle is the largest and strongest chamber in your heart. The left ventricle's chamber walls are only about a half-inch thick, but they have enough force to push blood through the aortic valve and into your body.

The heart has four types of valves which regulate blood flow through the heart:

- The tricuspid valve regulates blood flow between the right atrium and right ventricle.
- The pulmonary valve controls blood flow from the right ventricle into the pulmonary arteries, which carry blood to your lungs to pick up oxygen.
- The mitral valve lets oxygen-rich blood from your lungs pass from the left atrium into the left ventricle.

- The aortic valve opens the way for oxygen-rich blood to pass from the left ventricle into the aorta, your body's largest artery, where it is delivered to the rest of your body, Figure (10.1)

Figure (10.1): The structure of the human heart

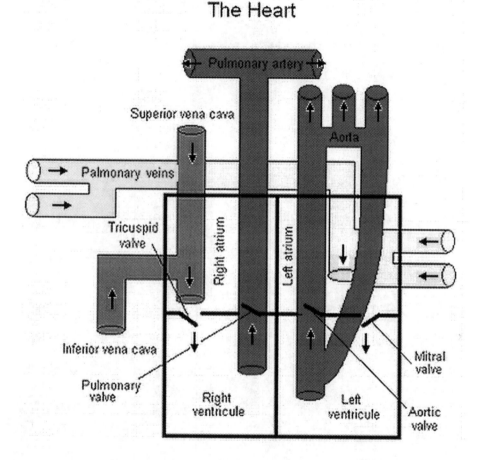

The Heart

10.3 Blood Vessels

If all the blood vessels including arteries, veins, and capillaries were laid end-to-end, they would extend for about 60,000 miles (more than 96,500 kilometers), which is far enough to circle the earth more than twice!

The heart works as a pump that pushes blood to the organs, tissues, and cells of the body. Blood delivers oxygen and nutrients to every cell and removes the carbon dioxide and waste products made by those cells. Blood is carried from your heart to the rest of your body through a complex network of arteries, arterioles, and capillaries. Blood is returned to your heart through venules and veins.

The heart itself, like every other organ or tissue in the body, needs oxygen-rich blood to work. The heart is equipped with a coronary circulation system that supplies the heart with oxygen and nutrients.

The contour of blood pressure, volume of the blood and the electrocardiogram (ECG) in the heart is shown in Figure (10.2).

Figure (10.2): Blood pressure, blood volume, and ECG in the heart (Wiggers Diagram)

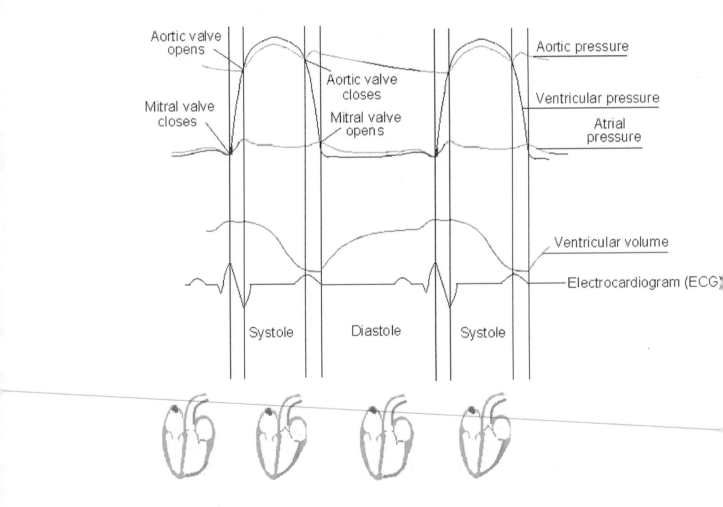

10.4 The Electrical System of the Heart

The rhythmic contractions and expansions of the heart that pump the blood occur in response to periodic electrical control pulse sequences.

Each beat of the heart is set in motion by an electrical signal from within the heart muscle. The signal starts from the SA node. This is why the SA node is sometimes called your heart's natural pacemaker. Your pulse (heart rate) is the number of signals the SA node produces per minute.

At rest, there is an electrical charge difference between the inside and outside of the neuron because of either positively or negatively charged ions that are caused by sodium (Na^+), potassium (K^+) and chloride (Cl^-). The inside of the neuron is more negatively charged than the outside of the neuron (because sodium is more than ten times more concentrated outside the neuron's membrane than inside of the neuron), and the neuron is said to be polarized, i.e., there is a difference in electrical charge between the inside and outside of the neuron. The neuron has channels that can permit chemicals to pass in and out of it. The sodium channels are completely closed during the resting potential, but the potassium channels are partly open, so potassium can flow slowly out of the neuron.

The protein of the neuron loves the potassium and hates the sodium, thus, the neuron pumps out the sodium and pumps in the potassium. Because the sodium atom ends with one electron and the chlorine in the sodium chloride (salt) ends with seven electrons, the sodium atom looses the electron to the chlorine atom. The sodium atom thenbecomes a positive ion (cation), and the chlorine atom becomes negative ion (anion), Figure (10.3).

Figure (10.3): Neurons and dynamical polarization

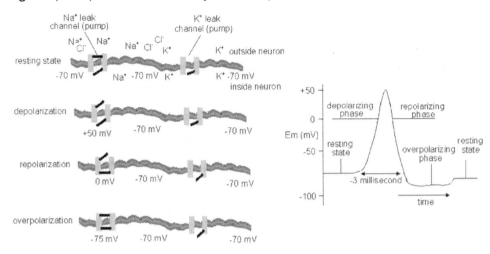

At rest, a cell has a potential (voltage) of -70mV. Once the membrane potential changes to -50mV, then the cell has been depolarized, i.e., the cell voltage becomes more positive. Depolarization is often caused by influx of cations, e.g. Na$^+$ through Na$^+$. the entrance of Ca^{++} through Ca^{++} channels can also cause depolarization. On the other hand, the efflux of the anion potassium (K$^-$) through K$^-$ channels inhibits depolarization, similar to the influx of Cl$^-$ through Cl$^-$ channels.

Electrical pulses are spontaneously generated by the SA node, the physiological pacemaker. These electrical pulses (one pulse at a time) are propagated throughout the right atrium, and then to the left atrium, stimulating the myocardium of both atria to contract. The conduction of the electrical pulse throughout the left and right atria is seen on the ECG as the P hump wave, Figure (10.4).

After a short delay (PR segment) the pulse goes from the SA node to the AV node. The time delay is necessary to avoid simultaneous contraction of both atriums and ventricles.

The pulses spread from the AV node into the His bundle, and then spread to the left and right bundles. The pulse in the left bundle is stronger than that of the right bundle because it has Purkinje fibers which have less resistance than normal fibers. The spread of electrical activity (depolarization) through the ventricular myocardium produces the QRS wave, Figure (10.4).

The last event of the cycle is the repolarization of the ventricles. The T wave represents the repolarization (or recovery) of the ventricles. The interval from the beginning of the QRS complex to the apex of the T wave is referred to as the absolute refractory period. The last half of the T wave is referred to as the relative refractory period or vulnerable period.The transthoracically measured PQRS portion of an electrocardiogram is chiefly influenced by the sympathetic nervous system which mobilizes the body's nervous system fight –or flight response. The T (and occasionally U) waves are chiefly influenced by the parasympahetic nervous system which is responsible for the regulation of internal organs and glands. This occurs unconsciously. Example are, salivation, urination, digestion, sexual arousal, etc.

An impulse that originates from the SA node at a relative rate of 60 - 100bpm is known as a normal sinus rhythm. If SA nodal impulses occur at a rate less than 60bpm (beat per minute), the heart rhythm becomes known as a bradycardiac sinus rhythm. If it occurs with a rate greater than 100bpm, it is called a tachycardiac sinus rhythm.

The sequence of the contraction of the heart chambers is:

1- An electrical pulse at SN expands the right atrium and opens the tricuspid valve. Blood comes to the right atrium through the superior/ inferior vena cava
2- The pulse goes from the SN to the AV node. Blood goes from the right atrium to the right ventricle through the tricuspid valve
3- The pulmonary valve opens to release blood to the lungs
4- Blood from lungs goes to left atrium
5- After the left atrium becomes full of blood, the mitral valve opens to let the blood gointo the left ventricle
6- After the left ventricle is full, the blood goes to the aorta through the aortic valve
7- Steps 3 to 6 are activated by the signals through the His bundle, the left bundle, the right bundle, and the purkinje fibers.

10.5 Cardiac Systole and Diastole

Systole means pressure, spasm or contraction. Two contractions occur in the heart; one in the atria and one in the ventricles. Atrial systole is the contraction of the heart muscle (myocardia) of the left and right atria. Normally, both atria contract at the same time. As the atria contract, the blood pressure in each atrium increases and will force additional blood into the ventricles.

Ventricular systole is the contraction of the muscles (myocardia) of the left and right ventricles. There is a time delay between the two ventricle systoles.

Cardiac Diastole is the period of time when the heart empties most of the blood into the aorta and the pulmonary artery. The heart then relaxes for a short period, after which it starts contraction. Ventricular diastole is when the ventricles are relaxing, while atrial diastole is when the atria are relaxing. Together they are known as complete cardiac diastole.

During ventricular diastole, the pressure in both left and right ventricles drops from its maximum pressure in the systole. The blood then starts to fill the ventricles though the pulmonary valve (from the right atrium), and through the mitral valve (from the left atrium).

Figure (10.4): Electrical conduction in the heart

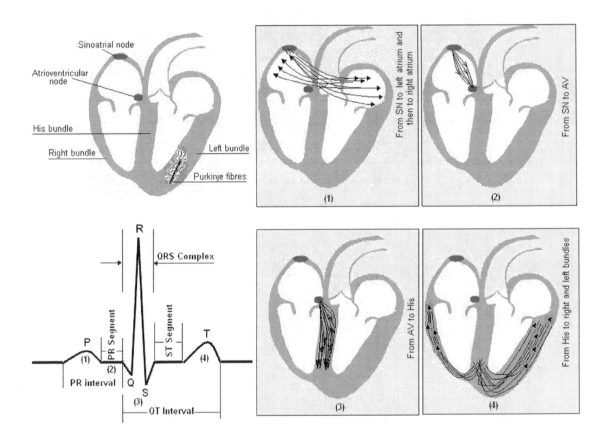

Chapter 11- The Electrical System of the Brain

The brain is represented by a closed box; of which the input can be one electrical pulse, and the output is of multi-order feed back loops. For example, you can measure the voltage, current and frequency of the electrical signals input to the brain, but you can not measure simultaneous individual signals distributed among the parts of the brain. Each individual signal passes though complicated pathways to execute certain functions such as perception, analysis, and task management. The voltage is determined primarily by the potassium and sodium ionic concentrations internal and external to the neuron, about -70mv at rest. Current flow would be ionic not electronic and is not measurable, "The Book of Intelligence and Brain Disorder" by this author.

Jeff Hawkins is an electrical engineer by training, and a neuroscientist by inclination. In his book (*How a New Understanding of the Brain will Lead to the Creation of Truly Intelligent Machines*) he used electrical engineering concepts as well as the studies of neuroscience to formulate his framework. In particular, Hawkins treats the propagation of nerve impulses in our nervous system as an encoding problem, specifically, a future predicting state machine, similar in principle to feed-forward error-correcting state machines.

The brain is extremely complex. The cerebral cortex of the human brain contains roughly 15–33 billion neurons. The cerebrum is the largest part of the brain, accounting for 85 percent of the organ's weight. The distinctive, deeply wrinkled outer surface is the cerebral cortex, which consists of gray matter. Beneath this lies the white matter.

The neurons are linked with up to 10,000 synaptic connections each. Each cubic millimeter of cerebral cortex contains roughly one billion synapses. These neurons communicate with one another by means of long protoplasmic fibers called axons, which carry trains of signal pulses called action potentials (measured in milli voltages and micro voltages) to distant parts of the brain or body and target them to specific recipient cells.

It is very hard to imagine how all the complex intellectual capabilities conferred by the human brain can develop from an embryo, which begins from a single cell, or rather the first two cells, that create it.

The human nervous system starts to form very early in the embryo's development. At the end of the gastrulation phase an elongated structure, the notochord, is laid down. The embryo thereby changes from a circular organization to an axial one.

In a mammalian embryo, the neural tube is initially a straight, linear structure as seen in Figure (11.1).

Figure (11.1): The brain of a mammalian embryo

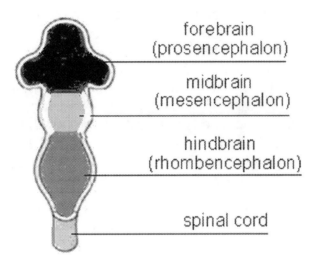

forebrain
(prosencephalon)

midbrain
(mesencephalon)

hindbrain
(rhombencephalon)

spinal cord

The brain produces electrical signals which, together with chemical reactions, let the parts of the body communicate. Nerves send these signals throughout the body.

11.1 The Structure of the Brain

The brain has two cerebral hemispheres. Each takes care of one side of the body, but the controls are crossed: the right hemisphere takes care of the left side, and vice versa, Figure (11.2). Each hemisphere appears to be specialized for some behaviors. For example, it appears that the right side of the brain is dominant for spatial abilities, face recognition, visual imagery and music. The left side of the brain may be more dominant for calculations, math and logical abilities. Of course, these are generalizations. In normal people, the two hemispheres work together, are connected, and share information through the corpus callosum. Much of what we know about the right and left hemispheres comes from studies in people who have had the corpus callosum split. When this happens a surgical operation isolates most of the right hemisphere from the left hemisphere. This type of surgery is performed in patients suffering from epilepsy. The corpus callosum is cut to prevent the spread of the "epileptic seizure" from one hemisphere to the other. The hemispheres communicate with each other through a thick band of 200-250 million nerve fibers called the corpus callosum. A smaller band of nerve fibers called the anterior commissure also connect parts of the cerebral hemispheres.

Figure (11.2): Cerebral hemispheres of the brain

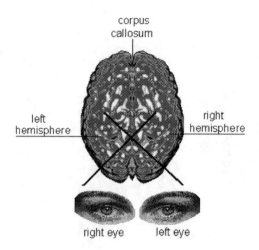

The brain is divided into three parts, as follows, Figure (11.3):

1) Forebrain: the telencephalon and diencephalon
2) Midbrain: the mesencephalon
3) Hindbrain: the metencephalon and myelencephalon

Figure (11.3): Parts of the brain

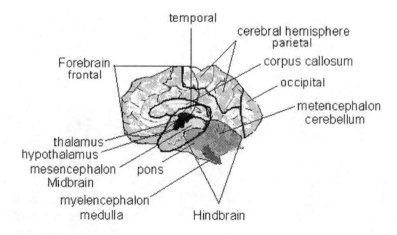

11.2 Functions of the Brain

There are several different regions of the human brain, each with interacting but distinct functions. Table (11.1) shows the regions with their functions.

Table (11.1): Regions of the brain and their functions

Frontal Lobe (Conscious Brain)	Consists of:
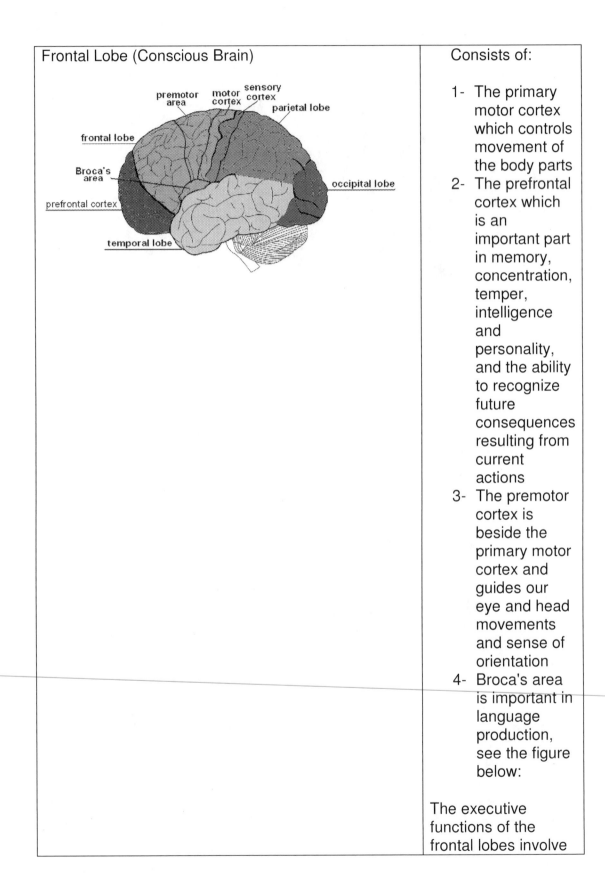	1- The primary motor cortex which controls movement of the body parts 2- The prefrontal cortex which is an important part in memory, concentration, temper, intelligence and personality, and the ability to recognize future consequences resulting from current actions 3- The premotor cortex is beside the primary motor cortex and guides our eye and head movements and sense of orientation 4- Broca's area is important in language production, see the figure below: The executive functions of the frontal lobes involve

	the ability to recognize future consequences resulting from current actions, to choose between good and bad actions, control voluntary movement, thinking, and feeling, override and suppress unacceptable social responses, and determine similarities and differences between things or events.
Parietal Lobe	The parietal lobe plays important roles in integrating sensory information from various parts of the body, knowledge of numbers and their relations, and in the manipulation of objects.
Occipital Lobe	o Contains the visual cortex o The visual cortex contains association areas that help in the visual recognition of shapes and colors. The right side of the brain 'sees' the left visual space, whereas the left side of the brain 'sees' the

	right visual space The primary visual cortex manages vision and has a full map of the visual world.
Temporal Lobe	o contains the auditory cortex helps that receives signals from the ear and lets us hear sounds and associate meanings with sounds o the Wernicke's area is important for language, speech and meaning The temporal lobe is involved in auditory perception and is home to the primary auditory cortex. It is also important for the processing of semantics in speech, hearing and vision. It contains the hippocampus and plays a key role in the formation of long-term memory.
Prefrontal Cortex	It is responsible for planning complex cognitive behaviors, personality expression, decision making and

	moderating correct social behavior. It helps focus attention, and gives meaning to perceptions.
Parietal Lobe	Parietal lobe coordinates signals received from other brain regions to interpret general sensory signals. It coordinates visual, auditory and language mechanisms , motor and sensory signals along with memory helps to identify objects
Thalamus	The major role of thalamus is to gate (relay) and otherwise modulate the flow of information (except smell) to cortex. For example, visual information from the retina is not sent directly to visual cortex but instead is relayed through the lateral geniculate nucleus of the thalamus.
Hypothalamus	One of the most important functions of the hypothalamus is to link the nervous system to the endocrine system via the pituitary gland.

In the Thalamus row, the left cell contains a labeled diagram with the labels: pineal gland, caudate nucleus, thalamus, amygdala, hippocampus, pituitary gland, hypothalamus.

	ITregulates basic biological drives, hormonal levels, sexual behavior, and controls autonomic functions such as hunger, thirst, and body temperature.
Pituitary Gland	It produces growth hormone, prolactin - to stimulate milk production after giving birth, ACTH (adrenocorticotropic hormone) - to stimulate the adrenal glands, TSH (thyroid-stimulating hormone) - to stimulate the thyroid gland, FSH (follicle-stimulating hormone) - to stimulate the ovaries and testes, and LH (luteinizing hormone) - to stimulate the ovaries and testes.
Hippocampus	It helps regulate emotion and memory. Functionally, the hippocampus is part of the olfactory cortex, that part of the cerebral cortex essential to the sense of smell. It mediates learning and memory formation.
Amygdala	It is a complex structure involved in a wide range of normal behavioral functions and psychiatric

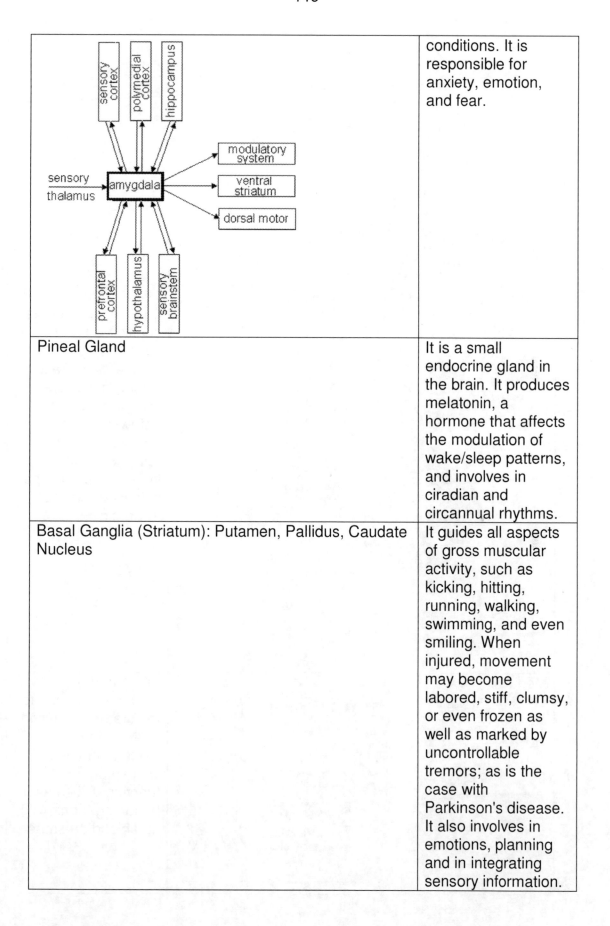	conditions. It is responsible for anxiety, emotion, and fear.
Pineal Gland	It is a small endocrine gland in the brain. It produces melatonin, a hormone that affects the modulation of wake/sleep patterns, and involves in ciradian and circannual rhythms.
Basal Ganglia (Striatum): Putamen, Pallidus, Caudate Nucleus	It guides all aspects of gross muscular activity, such as kicking, hitting, running, walking, swimming, and even smiling. When injured, movement may become labored, stiff, clumsy, or even frozen as well as marked by uncontrollable tremors; as is the case with Parkinson's disease. It also involves in emotions, planning and in integrating sensory information.

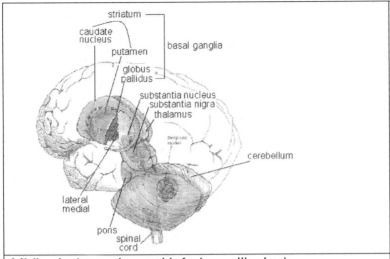

Midbrain (superior and inferior colliculus)	The midbrain (mesencephalon) is considered part of the brain stem. Its substantia nigra is closely associated with motor system pathways of the basal ganglia. It relays sensory information from the spinal cord to the forebrain.
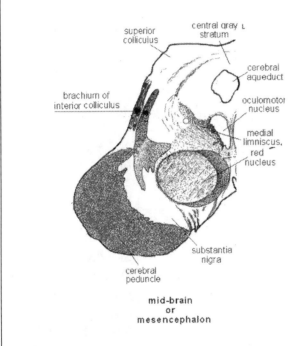	
Medulla 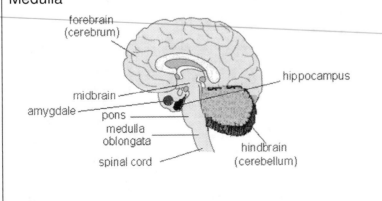	The medulla contains the cardiac, respiratory, vomiting and vasomotor centers and deals with autonomic functions, such as breathing, heart rate and blood pressure.

midbrain — eye medulla — lacrimal gland mucous memb., nose and palate submaxillar gland sublingual gland mucous membrane, mouth parotid gland heart larynx, trachea, bronchi esophagus stomach blood vessels liver pancreas small intestine large intestine	
Pons	The pones is located above the medulla. It contains nuclei that relay signals from the cerebrum to the cerebellum, along with nuclei that deal primarily with sleep, respiration, swallowing, bladder control, hearing, equilibrium, taste, eye movement, facial expressions, facial sensation, and posture. The nucleus of the pones regulates the change from inspiration to expiration.
Cerebellum	The cerebellum is involved in the coordination of voluntary motor movement, balance and equilibrium and

	muscle tone. The cerebellum does not initiate movement, but it contributes to coordination, precision, and accurate timing. It receives input from sensory systems and from other parts of the brain and spinal cord, and integrates these inputs to fine tune motor activity.
Limbic System	The limbic system is a complex set of structures that lies on both sides of the thalamus, just under the cerebrum. It includes the hypothalamus, the hippocampus, the amygdala, and several others (cingulate gyrus, ventral tegmental area, basal ganglia, prefrontal cortex) nearby areas. It appears to be primarily responsible for our emotional life, and has a lot to do with the formation of memories.
Cingulate Gyrus	It is an integral part of the limbic system, which is involved with emotion formation and processing, learning, and memory, and is also important for executive function and respiratory

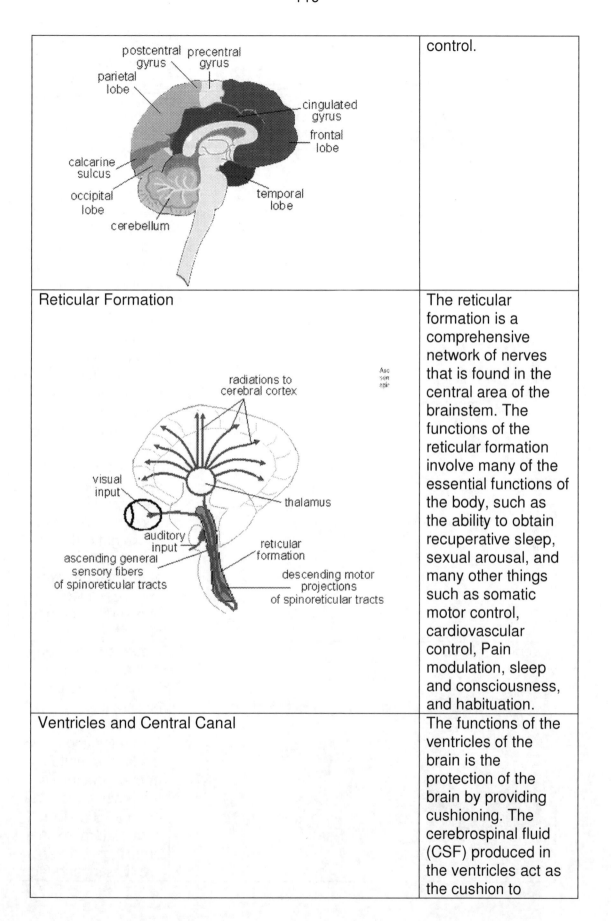

control.

| **Reticular Formation** | The reticular formation is a comprehensive network of nerves that is found in the central area of the brainstem. The functions of the reticular formation involve many of the essential functions of the body, such as the ability to obtain recuperative sleep, sexual arousal, and many other things such as somatic motor control, cardiovascular control, Pain modulation, sleep and consciousness, and habituation. |
| **Ventricles and Central Canal** | The functions of the ventricles of the brain is the protection of the brain by providing cushioning. The cerebrospinal fluid (CSF) produced in the ventricles act as the cushion to |

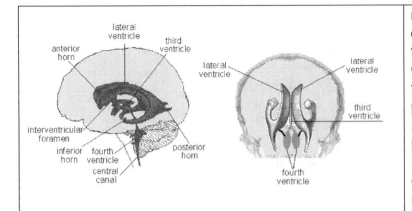

minimize the impact of any kind of trauma to the head. The CSF travels through the ventricles system provides a pathway and provides protection to the brain. CSF is concerned with the excretion of waste products such as, harmful metabolites or drugs from the brain, besides transporting the hormones to various part of the brain. It also provides buoyancy to the brain, which in turn, helps to reduce the weight of the brain. Just because our brain remains immersed in cerebrospinal fluid, its weight reduces from 1,400 gm to almost 50 gm, which in turn, reduces pressure at the base of the brain.

Corpus Callosum	It connects the left and right cerebral hemispheres and facilitates interhemispheric communication. It is the largest ehite matter structure in the brain, consisting of 200-250 million contralateral axonal projections. It involves in language

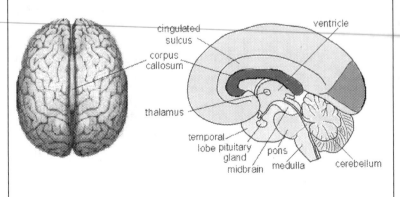

	learning.
Optic Chiasm 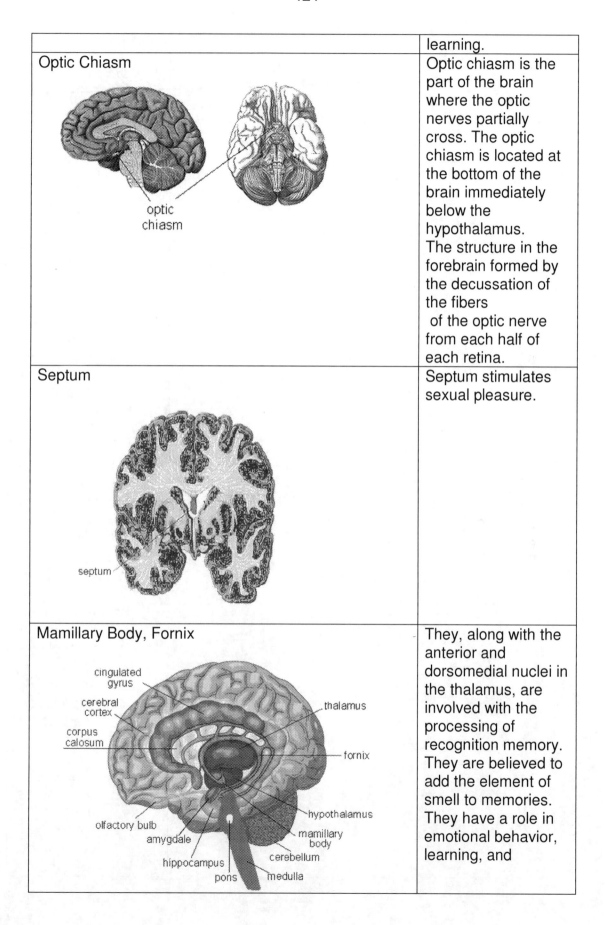	Optic chiasm is the part of the brain where the optic nerves partially cross. The optic chiasm is located at the bottom of the brain immediately below the hypothalamus. The structure in the forebrain formed by the decussation of the fibers of the optic nerve from each half of each retina.
Septum	Septum stimulates sexual pleasure.
Mamillary Body, Fornix	They, along with the anterior and dorsomedial nuclei in the thalamus, are involved with the processing of recognition memory. They are believed to add the element of smell to memories. They have a role in emotional behavior, learning, and

	motivation.
Spinal Cord	o The spinal cord receives information from skin, joints, and muscles o Sends back signals for both voluntary and reflex movements o Ttransmits signals from internal organs to the brain and from the brain to internal organs o Connects the brain to peripheral organs and tissue o Iin addition, the spinal cord contains • Ascending pathways through which sensory information reaches the brain • Descending pathways that relay motor commands from the

	brain to motor neurons
Functions of Brain's Parts 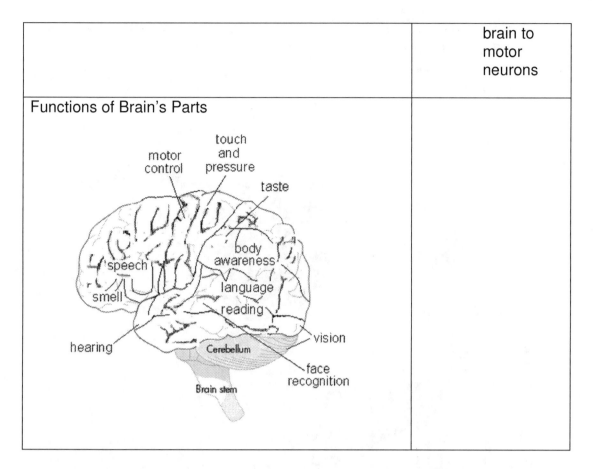	

Electrical signals from the brain's parts are carried out through neurotransmitters, which are endogenous chemicals that transmit signals from a neuron to a target cell across a synapse.

11.3 Neurons and Nerves

The human nervous system consists of two main systems: the central nervous system (CNS), and the peripheral nervous system (PNS), which includes the somatic motor nervous system, and the sensory nervous system. The central nervous system contains the majority of the nervous system and consists of the brain and the spinal cord, as well as the retina. The main function of the PNS is to connect the CNS to the limbs and organs. The peripheral nervous system is divided into the somatic nervous system and the autonomic nervous system.

11.3.1 The Neuron

The neuron consists of two portions: the cell body and the axon. The cell body is like the other cells. It contains a nucleus and cytoplasm. It is different from other cells because out of the cell body, long threadlike projections protrude. These are called dendrites ("tree" in Greek). At one point of the cell, however, there is a particularly long extension that usually does not branch throughout most of its sometimes enormous length. This is the axon (the axis), Figure (11.4).

Figure (11.4): A neuron

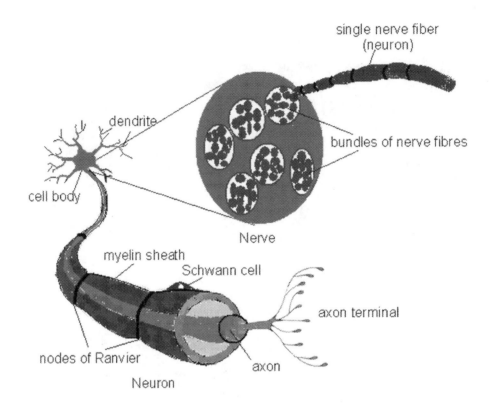

Structurally, a neuron consists of dendrites, anaxon and an axon terminal. Dendrites conduct nerve impulses toward the cell body. The axon conducts nerve impulses away from the cell body through the axon terminal. To speed up the transmission and to keep the signal from scattering and propagation (like an electrical cable), the axon is sheathed with a myelin layer that is made up of Schwann cells. Messages move as fast as 400 km per hour.

The axon terminal receives messages from the cell body, and then transmits the messages to neighboring neurons via the release of neurotransmitters. Neurotransmitters are endogenous chemicals that relay, amplify, and modulate signals between neurons and other cells, Figure (11.5).

Figure (11.5): Sending and receiving messages through neurons

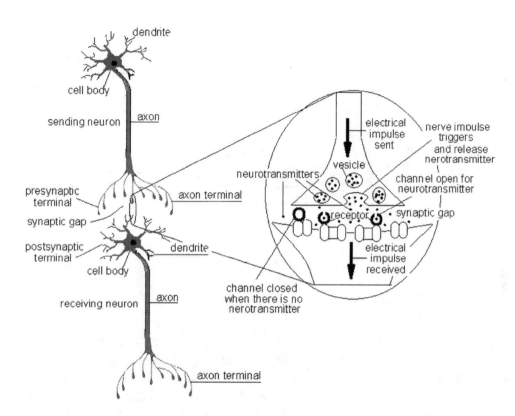

11.3.2 Types of Neurons

The three basic types of neurons are the motor neuron (efferent), the sensory neuron (afferent), and the interneuron. The motor neurons are specified to send messages away from the Central Nervous System. The sensory neurons are specified in the senses of taste, touch, hearing, smell, and sight. They send messages from the sensory receptors to the Central Nervous System. The interneurons are sort of a mix of both a sensory neuron and a motor neuron, Figure (11.6).

Figure (11.6): Sending messages to and from the Central Nervous System

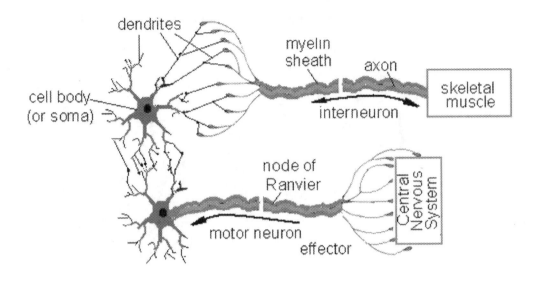

11.3.3 Transmitting Messages into the Neuron

At rest, there is an electrical charge difference between the inside and outside of the neuron because of either positively or negatively charged ions that are caused by sodium (Na^+), potassium (K^+) and chloride (Cl^-). The inside of the neuron is more negatively charged than the outside of the neuron (because sodium is more than ten times more concentrated outside the neuron's membrane than inside of the neuron), and the neuron is said to be polarized, i.e., there is a difference in electrical charge between the inside and outside of the neuron.

The neuron has channels that can permit chemicals to pass into and out of it. The sodium channels are completely closed during the resting potential, but the potassium channels are partly open, so potassium can flow slowly out of the neuron, Figure (11.7) repeated.

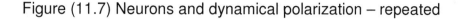

Figure (11.7) Neurons and dynamical polarization – repeated

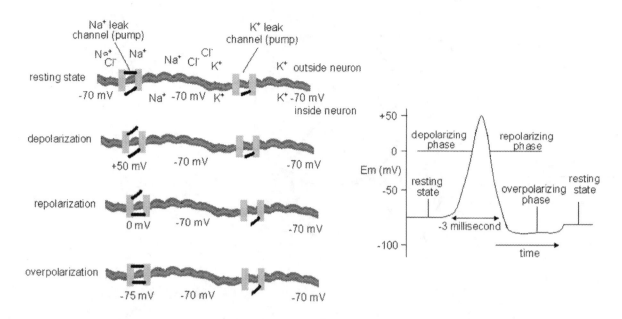

Substances that act as neurotransmitters can be broadly classified into three major groups:

1. Amino acids such as glumatic acid (glutamate), GABA, aspartic acid and glycine
2. Peptides such as vasopression, somatostatin, and neurotension
3. Monoamines such as neropinophrene, dopamine serotonin and acetylcholine

The central machines of the brain's neurotransmitters are glutamate and GABA. Some examples of neurotransmitter actions are:

a.   Dopamine – voluntary movement
b.   Serotonin – sleep and temperature regulation
c.   GABA (gamma aminobutryic acid) – motor behaviour
d.   Glycine – spinal reflexes and motor behavoiur
e.   Noradrenaline – wakeful and arousal
f.   Acetycholine – voluntary movement of the muscles
g.   Neuromodulator – sensory transmission (pain)
h.   Enkephalin (opiate) – stress, pain killer, promote calm
i.   ATP – energy
j.   Insulin - sugar

Figure (11.8) shows the effects on the mental states induced by three major neurotransmitters.

Figure (11.8): Effects of dopamine, serotonin and noradrenaline onto the mental states of the brain

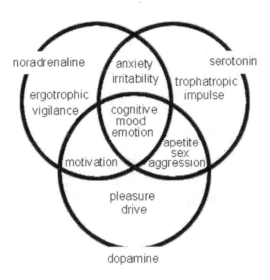

dopamine

11.3.4 Representation of the Brain's Signals

The author of this book (Dr. Amin Elsersawi) believes that incoming and outgoing signals from the basal ganglia to each part of the brain are similar to a spider web of multi-feed back loops. If you knock out any single "thread" of the web the rest of it works. He agreed with Swanson when she said, "there are usually alternate pathways through the nervous system. It's very hard to say that any one part is absolutely essential".

Dr. Elsersawi suggests an imaginative electrical diagram representing the brain network, as depicted in Figure (11.9).

Figure (11.9): Signal pathways in the brain (repeated)

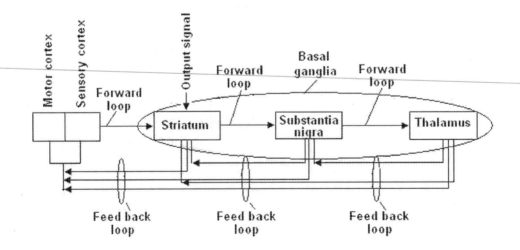

11.3.5 Brain Maps and the Secret to Intelligence (a Higher IQ)

1. New research suggests that the layer of insulation coating the neural wiring in the brain (myelin) plays a critical role in determining intelligence. In addition, the quality of this insulation appears largely to be genetically determined, providing further support for the idea that IQ is partly inherited. The neural wires that transmit electrical messages from cell to cell in the brain are coated with a fatty layer called myelin. Much like the insulation on an electrical wire, myelin stops currents from dissipating out of the wire and keeps the voltage level constant with which messages travel through the brain--the higher quality the myelin, the faster the messages travel. These myelin-coated tracts make up the brain's white matter, while the bodies of neural cells are called grey matter.

2. Does white matter play a key role in intelligence? The size of the corpus callosum, the thick tracts of white matter connecting the two hemispheres of the brain, is about 95 percent genetic. About 85 percent of the white-matter variation in the parietal lobes, which are involved in logic and visual-spatial skills, can be attributed to genetics (according to the neurologist Thompson at UCLA). But only about 45 percent of the variation in the temporal lobes, which play a central role in learning and memory, appears to be inherited. Thus, it's likely that the quality of white matter is at least partly genetically determined and, therefore, difficult to change.

3. Researchers have yet to find a simple neural explanation for intelligence. In 2001, Thompson showed that it is correlated with volume in the frontal cortex, a result consistent with a number of studies that have linked intelligence to overall brain size. But size is a crude measure: while larger brains may be smarter on average, it's not clear if that's because they have more nerve cells, more connections between cells, more of the fibers that carry neural signals, or less nodes that connect messages.

4. Scientists at UC Santa Barbara have made a major discovery in how the brain encodes memories. The team of scientists is the first to uncover a central process in encoding memories that occurs at the level of the synapse (where neurons connect with each other). Strengthening the synapse is a very important part of learning and memory.

Chapter 12- The Effects of Time-Varying Electromagnetic Fields on Physiological and Molecular Genetic of Human Neuronal Cells

Many methods have recently been engaged in an attempt to regenerate neural tissues. Bergen et al (1997) have used DC electric fields externally and internally to regenerate nerve cells in a variety of animal tissues.

The current studies are a combination of physiology and electromagnetic bioengineering, which relate generally to the fields of biophysics, tissue regeneration, tissue culture, and neurobiology. The current search relates to the use of a time-varying electromagnetic field for potentiating the growth of the cells and tissues of vertebrate animals. The search uses two-dimensional conducting plate electrodes and maybe three-dimensional tissue cultures. The NASA RWV tissue culture technologies have extended this three-dimensional capacity for a number of tissues and have allowed the tissue to express different genes and biomolecules.

Neuronal tissue has been largely intractable under conventional culture conditions, using the induction of an AC time-varying electromagnetic field in the culture region. However, DC currents and microgravity are being used to simulate the nerve growth, such as that of neuronal tissue which comprises elongated nerve cells, elongated axons, dendrites, and nuclear areas. Thus, massages from and to the brain can be transmitted faster through the axons (Lelkes and Unsworth, 1997).

12.1 How to Elongate Human Neural Cells Experimentally

Human neuronal progenitor cells (progenitor cells are similar to stem cells, but have a tendency to differentiate into a specific type of cell) are collected and put in a CO_2 incubator maintained at 37° and have a CO_2 concentration of 6%. Ciprofloxacin and Fungizone (for the protection against bacteria and fungus) are used to culture the cells.

Electrodes made of platinum and stainless steel are placed in the incubator, and connected to an AC pulsed generator. The generator produces waveform of 1-10 mA square waves, a 10 Hz variable and a modulated AC current, Figure (12.1).

Figure (12.1): Elongation of human neural cells

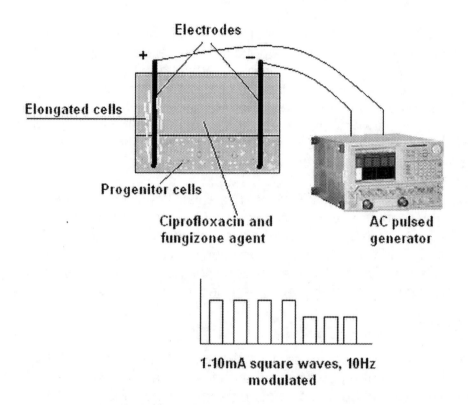

The elongation of neural cells should not use intense electric fields; weak and persistent stimulus (say at a frequency of 2-7 Hz and an intensity of the magnetic field of 7.5 pico Tesla) might yield better results.

Chapter13 - Magnetic Resonance

Magnetic resonance, in physics and chemistry, is a phenomenon produced by simultaneously applying a steady magnetic field and electromagnetic radiation (usually radio waves) to a sample of atoms and then adjusting the frequency of the radiation and the strength of the magnetic field to produce absorption of the radiation. The resonance refers to the enhancement of the absorption that occurs when the correct combination of field and frequency is reached. The procedure is analogous to tuning a radio dial exactly to a desired station

Several distinct kinds of magnetic resonance exist like cyclotron, Electron paramagnetic Resonance (EPR), Microscopic Magnets, and Nuclear Magnetic Resonance (NMR). In cyclotron resonance the magnetic field is adjusted so that the frequency of the revolution of a charged particle around the field lines is exactly equal to the frequency of the radiation. This principle is used to produce beams of energetic particles in particle accelerators.

13.1 The Cyclotron

A cyclotron is a particle accelerator in which charged particles in a static magnetic field are travelling outwards from the center along a spiral path and get accelerated by radio frequency electromagnetic fields, Figure (13.1).

The ion beams leaving the cyclotron can be used to treat cancer, similar to proton therapy. The ion beams penetrate the body at a certain energy to kill tumor cells, while minimizing damage to healthy cells. Cyclotrons are used to produce positive electrons (positrons) for Positron Emission Tomography (PET) imaging.

Cyclotrons are installed in hospitals for particle therapy, producing technetium-99. Technetium-99 (^{99}Tc) is an isotope of long –lived fission, which decays with a half-life of 211,000 years to stable ruthenium-99. The merit of Technetium in particle therapy is that it does not emit gamma rays; it emits only beta rays.

The cyclotron makes use of the magnetic force to bend moving charges into a semicircular path between accelerations by an applied electric field. The applied electric field accelerates electrons between the two "dees" of the magnetic field region. The field is reversed at the cyclone frequency to accelerate the electrons back across the gap. In a cyclotron, charged particles, such as protons, are produced at the center of the instrument. These particles move in a circular path because of the confining magnetic field above and below. The particles are alternately pushed and pulled by the alternating electric current, thereby acquiring more and more energy. As they move faster, they spiral outward. After about one hundred orbits they emerge from the instrument with great energy.

The cyclotron can accelerate a wide variety of ion species, from light ions like hydrogen and helium, to heavy ions like argon and xenon.

Figure (13.1): Cyclotron with its components

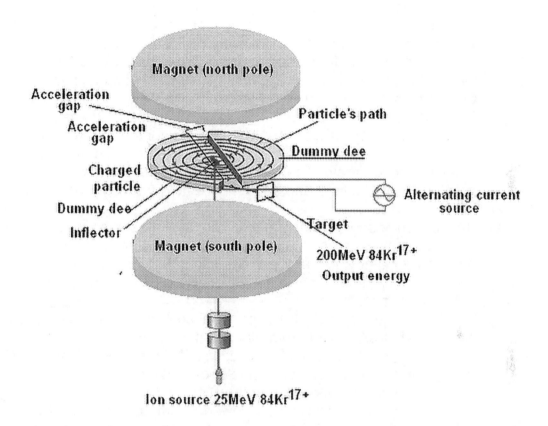

13.2 Magnetic Resonance Imaging (MRI)

The use of magnetic resonance imaging (MRI) scanners has grown tremendously. An MRI provides good contrast between the different soft tissues of the body, which makes it especially useful in imaging the brain, muscles, the heart, and cancers compared with other medical imagig techniques such as computed tomography (CT or X-rays). Unlike CT scans or traditional X-rays, an MRI does not use ionizing radiation.

An MRI scan is the best way to see inside the human body without cutting it open.

13.2.1 Components of the MRI Machine

Components of MRI machine are:

1) The tube: the patient is pushed into a tube that's only about 24 inches (60 centimeters) in diameter. This is the tube the patient enters, running through the magnet from front to back. The tube is known as the bore.

2) The main magnet: this is a very strong magnet. It is rated at 0.5 tesla to 2.0 tesla, or 5,000 to 20,000 gauss (1 tesla = 10,000 gauss). The earth magnetic field is about 0.5 gauss. The main magnet uses superconductive wires (this means no resistance or very small amount of ohm). The wires are continually bathed in liquid helium at 452.4 degrees below zero degrees Fahrenheit (269.1 below zero degrees Celsius). Therefore it is very expensive materials. The magnet creates an intense, stable magnetic field around the patient.

3) The resistive magnet: it does not require cryogens (such as liquid hydrogen or other refrigerant), but needs a constant power supply to maintain a homogenous magnetic field, and can be quite expensive to maintain. The resistive magnet provides a vertically oriented magnetic field and some fringe field for filling the opened flux return, if any.

4) The gradient magnets: they are three gradient magnets inside the MRI machine. They are much lower in strength compared to the main magnetic field; they may range in strength from 180 gauss to 270 gauss. They create a variable field, that allows different parts of the body to be scanned.

5) The Radiofrequency coils: these coils transmit radiofrequency waves into the patient's body. There are different coils for different parts of the body (knees, shoulders, wrists, heads, necks and so on). These coils usually conform to the contour of the body part being imaged, or at least reside very close to it during the scan.

6) The computer system: it is a very powerful computer system and a patient table, which slides the patient into the bore. The computer system uses Fourier analysis to create a picture from different dislocated hydrogen atoms.

13.2.2 How the MRI Machine Works

a. When hydrogen atoms of the patients (hydrogen atoms are abundant since the body is mostly made up of water and fat, and randomly spins in different directions during a steady state situation of the patient) are placed in a magnetic field, the atoms line up in the direction of the field. So they will be lined up horizontally, vertically, diagonally or as per the direction of the resultant of the magnetic fields.

b. These hydrogen atoms have a strong magnetic moment (because hydrogen is considered to be positive ions), which means that in a magnetic field they line up in the direction of the field. Since the magnetic field runs straight down the center of the machine, the hydrogen protons line up parallel to the patient. This is true at the beginning of the scan. The protons cancel each other, i.e. protons lined up toward the feet cancel those lined up toward the head of the patient. Only a couple of protons out of every million aren't canceled out. If the number of hydrogen atoms in the body is estimated by 7×10^{27}, this means that about 14×10^{25} hydrogen atoms are not cancelled, and there are enough to create extremely detailed images.

c. During the scanning, the radio frequency pulses are applied toward the area of the body required to scan. When the pulses are applied, the unmatched protons of hydrogen absorb the energy and spin again in a different direction. This is the "resonance" part of an MRI, and it is called the Larmour frequency. This resonance frequency will be calculated by the super computer attached to the MRI machine, using Fourier analysis.

d. The three gradient magnets start to act at the same time. When they are switched on and off, the hydrogen protons reverse their direction. The reversal of their direction produces energy that is calculated by the computer. Figure (13.2) illustrates the components of an MRI machine.

Figure (13.2): Components of MRI machine

13.3 Computed Tomography (CT)

A C.T. machine is an X-ray scanner of cross sectional images of body parts. It takes a series of special X-ray pictures of a region of the body arranged as a set of image "slices". The image slices are obtained from the top to the bottom of the region of interest and each slice contains information about the organ, Figure (13.3).

C.T. scans provide highly detailed images of every region of the body. CT scanners are now widely used for diagnostic medical imaging and security screening. They are most commonly used to examine the brain, chest, abdomen/pelvis and back. A C.T. is useful for evaluating the body for signs of infection, masses, trauma, degenerative changes etc. Often, C.T. scanning is the only way this information can be obtained. C.T. examinations sometimes require the use of injected contrast material (X-ray dye), which could affect the kidney. Your doctor must take blood work after the C.T. scan to ensure that the kidney is working properly.

In today's state-of-the-art medical and security CT scanners, over 1,000 images are collected in less than one second by a high-speed rotation of an x-ray tube around the object. Multiplexing is a process of combining multiple signals to form one composite signal for transmission. Since multiplexing involves transmitting many signals simultaneously, a high speed computer is required. Limitations of C.T. scanners are the modulating and rapidly changing images, and the cooling of the metal filament of the X-ray tubes.

There are two types of C.T. scan machines: one has only one X-ray tube and the other has multi X-ray tubes, as shown in Figure (13.3).

Figure (13.3): A single and multiple C.T. scan machine with components

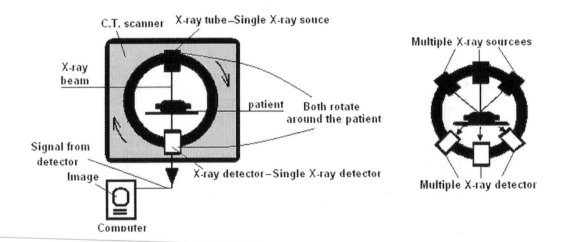

C.T. scan has less contrast compared to an MRI, and can pose the risk of radiation. It uses ionizing radiation. Images are more detailed compared to an MRI; however an MRI is more versatile than the C.T. scan and is used to examine a large variety of medical conditions. MRI images are more expensive and slower in time than a C.T. scan.

GLOSSARIES

Alpha particle is a particle consisting of two protons plus two neutrons. It is emitted by a radionuclide.

Alternating Current (AC) is a type of electrical current, in which the direction of the flow of electrons switches back and forth at regular intervals or cycles.

Atmosphere consists of nitrogen (78%), oxygen (21%), and traces of other gases such as argon, helium, carbon dioxide, and ozone.

The atmosphere plays an important role in the protection of life on Earth; it absorbs ultraviolet solar radiation and reduces temperature extremes between day and night.

Atom is the smallest portion of an element that can combine chemically with other atoms.

Beta particle is an electron emitted by the nucleus of a radionuclide. The electric charge may be positive, in which case the beta particle is called a positron.

Cells of higher organisms (known as eukaryotes) are subdivided into subcellular compartments called organelles such as the mitochondrion, the cell nucleus, the endoplasmatic reticulum, the Golgi apparatus and many smaller organelles with highly specialized functions.

Cell Membrane is the cell membrane (also called the plasma membrane) is the interface between the cellular machinery inside the cell and the fluid outside. In essence membranes are essential for the integrity and function of the cell.

Cellular Metabolism is the entire sum of all physical and chemical processes is constantly taking place in human cells. Healthy cells need food, oxygen, water and nutrients, which produce energy for the functioning of the human body. At the same time, waste and toxins are released.

Cosmic rays are high energy ionizing radiations from outer space. There is lot of complex composition at the surface of the earth.

Current is the flow of electricity commonly measured in amperes.

DNA constitutes the molecules inside cells that carry genetic information and pass it from one generation to the next.

Diagnostic radiology is a term usually applied to the use of X-ray in medicine for identifying disease or injury in patients.

Direct current (**DC**) is the unidirectional flow of electric charge. Direct current is produced by sources such as batteries, thermocouples, solar cells, and commutator-type electric machines of the dynamo type.

DNA is called "Deoxyribonucleic acid." It is the compound that controls the structure and function of cells and is the material of inheritance.

EEG is an electroencephalography (EEG), in the broadest sense of the term, refers to the measurement of the electrical activity produced by the brain.

Electric field is an invisible force field created by the attraction and repulsion of electrical charges, and is measured in Volts per meter (V/m). Electric Field is the region around any electrically charged material contains an electric field that affects other charged objects. The field around a negatively charged material pulls positively-charged objects in toward the material, while negatively-charged objects are pushed away. Around a positively charged material, on the contrary, negative objects are attracted and positive objects are repelled. The strength of the field gets rapidly weaker as one move further away from the charged material.

Electromagnetic field (EMF) is an invisible zone of energy that surrounds electric devices and wiring. EMFs are comprised of two fields: an 'electric' field and a 'magnetic' field. The electric field is created by voltage, which determines the *force* with which the electricity is pushed through wires. The magnetic field is created by the current, which is the amount of electricity being pushed. There are two types of electromagnetic fields:

Electromagnetic radiation is produced by radiation on an atomic scale, such as electrons in atoms. EM radiation has an electric and magnetic field component which oscillate in phase perpendicular to each other and to the direction of energy propagation. Electromagnetic radiation is classified into types according to the frequency of the wave; these types include (in order of increasing frequency): radio waves, microwaves, infrared, visible light, ultraviolet, X-rays and gamma-rays. Of these, radio waves have the longest wavelengths and gamma rays have the shortest.

Electrons are particles that carry negative electrical charges. Electrons form the outer "reactive" shell of atoms which interacts with other atoms and form the chemical bonds that hold molecules together. Flow of electrons between two points generates an electric current. Electron is one of the particles that make up an atom and particles that, in motion, can form an electric current. An electron is 2000 times lighter than the lightest atom.

Electrosmog is a New form of pollution coming from unnatural man-made EMF sources, such as power lines, microwaves, hair dryers, computers, etc. that pose harm to people's health. Though these assaults are subtle on an individual basis, these days we are continuously surrounded by these unnatural man-made EMFs and we are not aware that this level of unprotected exposure can cause significant health risks.

Energy is the capacity for, or the ability to do, mechanical work. Electrical energy is measured in kilowatt-hours for billing purposes.

Frequency is the number of complete alternations or cycles per second of an alternating current. It is measured in Hertz. The standard frequency in the US is 60 Hz. However, in some other countries the standard is 50 Hz

Hertz is a number of cycles per second. It is a measurement term for quantifying electro/pulsed magnetic frequencies.

Ion is an atom or molecule which has lost or gained one or more electrons, giving it a positive or negative electrical charge. A positive charged ion has fewer electrons than protons and negative charged ion has more electrons. Ions are essential to life; Sodium, Potassium, Calcium and other ions play an important part in cells of living organism, particularly in cell membranes. Ion is a positively or negatively charged atom or molecule.

Isotope is a nuclide with the same number of protons but different numbers of neutrons.

Fibromyalgia is a chronic pain disorder characterized by widespread pain of the muscles and bones, stiffness, general fatigue, and sleep disturbances. The underlying cause remains unknown, yet most researchers agree that it is related to the nervous system. There are several suggested explanations for fibromyalgia, such as genetic predisposition, stress, trauma, psychological problems. Treatment includes pain and sleep management and psychological support.

Frequency is the measurement of the number of times that a repeated event occurs per unit of time. The frequency of wave-like patterns including sound, electromagnetic waves (such as radio or light), electrical signals, or other waves, expresses the number of cycles of the repetitive waveform per second.

Ground is a conducting connection between an electrical circuit or device and the earth. A ground may be intentional, such as in the case of a safety ground, or accidental which may result in high overcurrents.

Magnetic field is an invisible force field created by a magnet or as a consequence of the movement of electric charges (flow of electricity). The magnitude (intensity) of a magnetic field is usually measured **Tesla** (T or in mT), but it can also be measured in **Gauss** (G).

Man-made electromagnetic fields (EMF) are for example generated by extremely low frequency (ELF) sources, such as power-lines, wiring and appliances as well as by higher frequency sources such as radio and television waves and, more recently, cellular telephones and their antennas.

Naturally occurring EMF are for example, the earth static magnetic field to which we are constantly exposed, electric fields caused by electrical charges in the clouds or by the static electricity produced when two objects are rubbed together as well as sudden electric and magnetic fields caused by lightning, etc.

Ohm is a unit of electrical resistance. A circuit resistance of one ohm will pass a current of one ampere with a potential difference of one volt. It is abbreviated using the Greek letter omega (Ω), and named for the German physicist George Simon Ohm 1854.

Photon is a quantum of electromagnetic radiation.

Protein is a large molecule composed of one or more chains of amino acids in a specific order, formed according to genetic information.

Radiation dose is a General term for quantity of ionizing radiation, used for treatment.

Resonance is a vibration of large amplitude in a mechanical or electrical system caused by a relatively small periodic stimulus of the same or nearly the same period as the natural vibration period of the system: the state of adjustment that produces resonance in a mechanical or electrical system. The classic example is demonstrated with two similar tuning-forks of which one is mounted on a wooden box. If the other one is struck and then placed on the box, then muted, the un-struck mounted fork will be heard. Human cells each have a specific vibration. The Medithera produces an electromagnetic field, which imitates and creates the same vibration as healthy human cells.

Transmembrane Potential is the electric voltage of the outer layer of human cells. In a healthy cellular environment the transmembrane potential is between 70-110 mV (millivolts). In a state of disease the cells of the human body drop their voltage under 70 mV, thus causing a sluggish metabolism. The Medithera assists in raising and maintaining the cell transmembrane potential to 70-110 mV.

Ultraviolet radiation is an electromagnetic radiation of shorter wavelength than visible light but of longer wavelength than x-rays, i.e. ranging from approximately 400 nm to 100 nm. The most common source of ultraviolet radiation is the sun, but it can also be produced artificially by UV lamps.

Vertigo is a dizzying sensation of the environment spinning, often accompanied by nausea and vomiting.

Volt is the electrical potential difference or pressure across a one ohm resistance carrying a current of one ampere. It is named after Italian physicist Count Alessandro Volta 1745-1827.